P9-DMD-145

The Insect Cookbook

ARTS AND TRADITIONS OF THE TABLE

The Insect Cookbook

Food for a Sustainable Planet

Arnold van Huis, Henk van Gurp,
and Marcel Dicke

Translated by Françoise Takken-Kaminker
and Diane Blumenfeld-Schaap

Columbia University Press   New York

Columbia University Press
Publishers Since 1893
New York   Chichester, West Sussex

cup.columbia.edu

Het Insectenkookboek © 2012 by Arnold van Huis, Henk van Gurp, and Marcel Dicke. Originally published by Uitgeverij Atlas Contact, Amsterdam
English-language edition copyright © 2014 Columbia University Press
All rights reserved

This book was made possible through support from the DOEN Foundation, the Uyttenboogaart-Eliasen Foundation, Wageningen University, and the Dutch Foundation for Literature.

Library of Congress Cataloging-in-Publication Data
Huis, Arnold van.
 [Insectenkookboek. English]
 The Insect cookbook : food for a sustainable planet / Arnold van Huis, Henk van Gurp, and Marcel Dicke ; translated by Françoise Takken-Kaminker and Diane Blumenfeld-Schaap. — English-language edition.
    pages cm — (Arts and traditions of the table : perspectives on culinary history)
 Originally published by Uitgeverij Atlas Contact, Amsterdam in 2012 as Het Insectenkookboek.
 Includes index.
 ISBN 978-0-231-16684-3 (cloth : acid-free paper)
 ISBN 978-0-231-53621-9 (e-book)
 1. Cooking (Insects) 2. Edible insects. I. Gurp, Henk van. II. Dicke, Marcel. III. Title.

TX746.H8513 2014
641.6'96—dc23

                                                                                    2013030373

Columbia University Press books are printed on permanent and durable acid-free paper.
This book is printed on paper with recycled content.
Printed in the United States of America

c 10 9 8 7 6 5 4 3 2 1

FRONTISPIECE: Tortilla with grasshoppers and guacamole. (Lotte Stekelenburg)
COVER IMAGES: Darlyne A. Murawski © Getty and *Cultura Limited / SuperStock*
COVER DESIGN: Evan Gaffney

References to Web sites (URLs) were accurate at the time of writing. Neither the authors nor Columbia University Press is responsible for URLs that may have expired or changed since the manuscript was prepared.

Contents

1 Insects: Essential and Delicious

2 Is It Healthy?

RECIPES: FIVE APPETIZERS

3   Eating Insects: Naturally!

JAN RUIG, POULTRY AND GAME SPECIALIST
"Some People Won't Try Anything New"   82

RECIPES: ELEVEN MAIN DISHES

DANIELLA MARTIN, EDIBLE-INSECT ADVOCATE
"Valuable, Abundant, and Available to Everybody"   108

ROBÈRT VAN BECKHOVEN, PASTRY CHEF
"Bonbon *Sauterelle*"   113

RECIPES: FIVE FESTIVE DISHES

RECIPES: SIX DESSERTS

4 On the Future and Sustainability

Migratory locust, reared in the Netherlands for culinary use. (Lotte Stekelenburg)

Foreword

The first time I ate an insect was somewhere in Burundi, in the 1980s. I was accompanying a farmer on her walk through an area that had been forested in the past, where now there were only old, poorly tended oil palm trees dating from the Belgian colonial days. One of the palm trees had been knocked down in a storm the previous night. When the farmer noticed this felled tree, she ran to it enthusiastically. Her eager fingers searched within the leaf axils just underneath the tree's crown, and she triumphantly pulled out a handful of fat white grubs. After wrapping the insects carefully in some leaves, she took them home and roasted them that same afternoon. I became curious at her evident delight in this find. And indeed, the small crispy treats were delicious! Shortly thereafter, I was offered honey-flavored, roasted grasshoppers in Syria . . . and was by then convinced: insects not only are a good idea nutritionally, but also taste wonderful.

Taste, however, is only one important aspect to bear in mind. Of prime importance to me, as an agricultural specialist, is that insects can contribute to protein supplies worldwide. I see the production of animal protein to feed a growing world population as one of the major scientific and social challenges of our times. Animal protein (from meat, fish, dairy products, eggs) is essential for everyone, but especially so for children, pregnant women, and the vulnerable elderly. Although a healthy adult who eats a large variety of foods can, for a time, do without animal protein, the diet of most people isn't adequately balanced.

People everywhere in the world have a preference for animal protein, except in areas where Hinduism is practiced. Consuming animal protein imparts status because it makes people stronger. This is a fact of evolution. The production of meat, dairy products, and fish, however, is problematic because of its high cost in land, water, and chemicals.

A large intake of animal protein is also directly linked to lifestyle-related diseases in wealthy countries. It is therefore important to curb the per capita consumption of meat and (especially fatty) dairy products where this intake is too high, while making animal proteins available in countries where their consumption is too low. In order to feed the 9 billion people who will probably populate the earth in 2050, we need to increase our production of animal proteins by 73 percent, a substantially greater amount than the increase needed in grain production. Greenhouse effects, as well, make it imperative that we search for alternatives.

Paleontological studies have shown that insects were included in the diet of early humans. Interestingly, they are also part of the Judeo-Christian culture. In the Bible, grasshoppers were recommended, on the one hand (in Leviticus), and forbidden, on the other (Deuteronomy). The fact that it was necessary to forbid them indicates that there was a long tradition of consuming insects. For food taboos, there is seldom a clear-cut medical explanation; rather, they are of a symbolic value. This is certainly evident in the obvious aversion to insects present today in the Western world. Insects are considered creepy. Strange, actually, because we eat animals out of the sea that closely resemble them, such as shrimp, lobster, and eels. But then we call those animals "seafood," and in other languages they are referred to as "sea fruits" as a way to obscure their animal origin.

The importance of insects lies not only in the fact that they are an alternative source of protein. Eating insects can change your perspective, allowing you to consider how you take your eating habits for granted. Nutrition not only is biology or history, but also has to do with customs and choice. Eating insects paves the way to eating many other animal species that are further down the food chain, such as snails and algae, which can also boost the global food supply.

This book highlights insects as a delicacy, as a special addition to meals. Insects can play an even more important role as "processed meat," such as that found in sausages, soups, sauces, or pizzas. Insects can replace a portion of the meat currently used for these products. I would prefer to see this done in combination with plant proteins, so that a sausage would consist of 30 percent traditional meat, 35 percent insects, and 35 percent plant protein. This would make a tremendous difference, as people all over the world, in particular because of urbanization, are consuming ever-increasing amounts of ready-made meals—especially those containing meat.

In a playful yet thorough manner, this cookbook effectively violates a taboo. It calls for more: not only because it is important for the future of our food, but also because eating insects makes us aware of our place in the food chain, aware of ancient traditions and blind habits as well as new possibilities. More than ever, we need to realize that food is not just a plastic container of cheap calories, but something that links us with the present and the past, and with the entire ecological cycle of life on Earth.

LOUISE O. FRESCO

UNIVERSITY PROFESSOR, UNIVERSITY OF AMSTERDAM

HONORARY PROFESSOR, WAGENINGEN UNIVERSITY

FORMER ASSISTANT DIRECTOR-GENERAL OF THE FOOD

AND AGRICULTURE ORGANIZATION OF THE UNITED NATIONS

Dishes containing insects in Laos. (Joost Van Itterbeeck)

Preface

Ever since we started to promote entomophagy (eating insects) in the Netherlands, we have received tremendous interest from the public at large. Not being accustomed to eating insects, people were curious about this new food source, in particular considering the major challenge of producing sufficient animal protein for a rapidly increasing population on our planet—expected to reach 9 billion people in 2050.

Because of the environmental benefits of insects as a new protein source, we often heard, "When do you expect that insects will be available in the supermarket?" This possibility became a reality in 2008, when Dutch companies started to produce insects for human consumption. Then, of course, the question was: How do you cook the insects? As a response, we produced *Het Insectenkookboek* (*The Insect Cookbook*) in Dutch. This book provides not only recipes, but also examples of insects being eaten all over the globe: which insects, where, and why. Moreover, the book contains interviews with chefs, politicians, and food designers. This English version of the book is updated and contains new interviews: with Kofi Annan, the former secretary-general of the United Nations, and René Redzepi, who is considered to be one of the best chefs in the world. The recipes can be made at home by any amateur cook interested in new cuisine. The insect ingredients are three species that are available through the Internet for human consumption:

- Migratory locusts (*Locusta migratoria*). Locusts are grasshopper species that under favorable circumstances develop into migratory and gregarious animals. The wingless larvae form marching bands, and the winged adults migrate in huge swarms.
- Yellow mealworms (*Tenebrio molitor*), larvae of flour beetles.
- Buffalo worms (*Alphitobius diaperinus*), also larvae of flour beetles, resembling the yellow mealworm but smaller.

The issue of entomophagy will be addressed during an international conference in May 2014, organized by Wageningen University in conjunction with the Food and Agriculture Organization of the United Nations. This will certainly not be the last development; it is likely that, in 2024, we will look back and wonder why insects were not yet frequently on the menu back in 2014. We hope that this book will contribute to that development. After all, feeding a rapidly growing human population is a serious issue.

Girls with strings of grasshoppers in Indonesia. (Annie Monard FAO)

Acknowledgments

Many people have contributed to the realization of this book, and we would like to sincerely thank Kofi Annan, Robèrt van Beckhoven, Margot Calis, Jan Fabre, Katja Gruijters, Harmke Klunder, Daniella Martin, Marian Peters, Edoardo Ramos Anaya, René Redzepi, Jan Ruig, Johan Verbon, Herman Wijffels, and Pierre Wind for their enthusiastic participation. We thank Louise O. Fresco for writing the foreword, and Paul Vantomme for placing the subject of our book in an international context. We are grateful to Françoise Takken-Kaminker and Diane Blumenfeld-Schaap for the English translation; and to Emile Brugman and Erna Staal at Atlas Contact for their confidence in this book. Lieke van Gurp we thank for her help with the recipe texts. We appreciate the Rijn IJssel Vakschool Wageningen for providing its location for photographing the dishes.

This book was made possible through support from the DOEN Foundation, the Uyttenboogaart-Eliasen Foundation, and Wageningen University.

For preproduction coordination of the manuscript, we thank Caroline Zeevat and Marian Peters. For the interview texts, we thank Maartje Berendsen (Robèrt van Beckhoven, Margot Calis) and Drees Koren (Jan Fabre, Katja Gruijters, Harmke Klunder, Marian Peters, Edoardo Ramos Anaya, Jan Ruig, Johan Verbon, Herman Wijffels, Pierre Wind). For portrait photography and interview photos, we thank Lotte Stekelenburg (except the portraits of Kofi Annan, Daniella Martin and René Redzepi; the photographers are credited in the captions). For photography of recipes (from Henk van Gurp), we thank Floris Scheplitz.

The Insect Cookbook

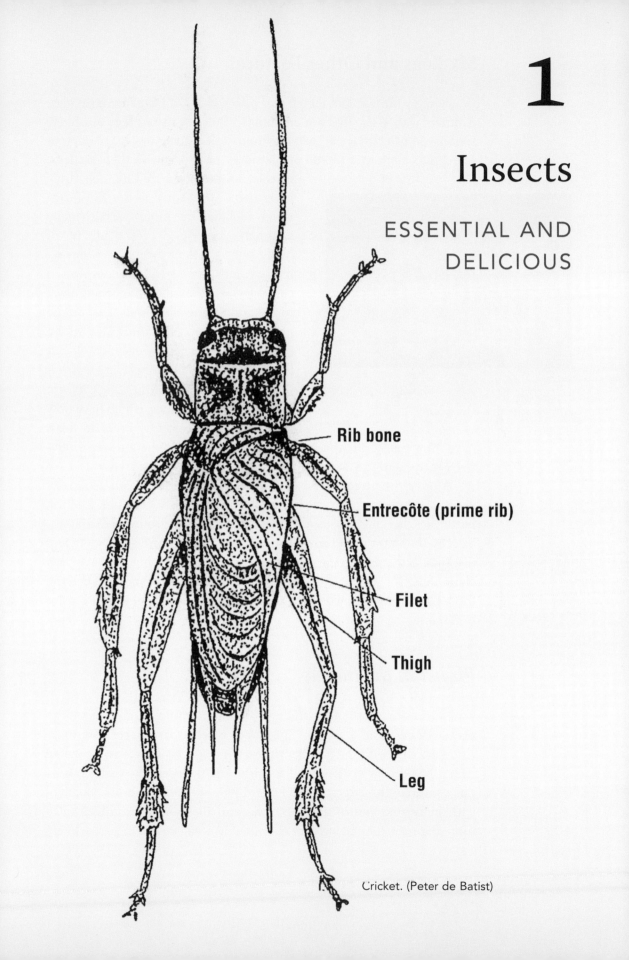

Rib bone

Entrecôte (prime rib)

Filet

Thigh

Leg

Cricket. (Peter de Batist)

Six Legs and Other Features

"Show me your legs and I'll tell you who you are." Insects have six legs. Animals with eight legs are arachnids. Those with ten legs are likely some type of crustacean, whereas four legs usually means they are a vertebrate, such as a reptile or mammal. In addition to their six legs, insects can be recognized by their three body parts. The front part is the head- which includes the mouthparts (usually external); compound eyes made up of many small parts or facets, each having its own lens; and two antennae (often called feelers) that allow the insect both to feel and to smell very well. The middle part of the insect, the thorax, is where not only the six legs are attached, but in adult insects, also the wings. Adult insects generally have four wings, except for flies and mosquitoes, which have only two. And some insects, such as fleas, have no wings at all. The insect abdomen contains the gut, reproductive organs, and the fat body, where food reserves are stored.

Honeybee. (Hans Smid [www.bugsinthepicture.com])

Among the insects, we can distinguish twenty-six main groups or orders, including, for example, mayflies, dragonflies, termites, beetles, butterflies, lice, and wasps and bees. Each of these groups is characterized by the main insect traits, but other features can vary widely and distinguish them from the other groups. For example, beetles have hard elytra (anterior wings), and wasps have a narrow waistline between the thorax and abdomen, whereas the three body parts of aphids are harder to discern.

Planet of the Insects

Insects are the most successful and abundant group of animals on the planet. More than 1 million species of insects have been described. Compare that with the number of mammal species known: only 5,400! About 80 percent of all animal species walk the earth on six legs. Beetles comprise the largest number of different species (about 360,000), with butterflies (more than 150,000) and flies and mosquitoes (also more than 150,000) being very abundant as well. New species are being discovered every day, in particular, but certainly not exclusively, in tropical rainforests. We estimate that there are about 6 million insect species on Earth. And that's not all—estimates have also been made of the total number of individual insects on the planet: between 10^{18} and 10^{19}, numbers we can't even imagine. This would mean that for every human being on Earth, there are between 200 million and 2 billion insects. If every insect were to weigh 1 milligram, which is the weight of a small ant, then for each one of us, there are 440 to 4,400 pounds (200–2,000 kg) of insects. In other words, insects dominate the world not only according to numbers of species, but also in terms of their total weight. Our planet is actually a planet of insects, on which a few people also live.

Variations on a Theme

The many species of insects make up a diverse palette. Some are smaller than 0.5 millimeter—smaller than the period at the end of this sentence—such as parasitic wasps that lay their eggs in other insects, or certain beetles. Other species, such as the Atlas butterfly, can reach up to 1 foot (30 cm) in wingspan. Indi-

Cockchafer, a large European beetle. (Hans Smid [www.bugsinthepicture.com])

vidual insects can vary in weight from 10 micrograms (0.01 mg) for a small wasp to as much as 3.5 ounces (100 g) for a Goliath beetle. Varieties in insect shapes and colors are also manifold: flies with beautifully colored eyes; butterflies having lovely patterned wings; long-necked insects, such as the giraffe beetle; and ants, with unbelievable differences in size between the soldier ants and the workers. Have a look at an insect collection and you will be filled with wonder at the enormous variety in color, shape, size, and lifestyle.

Insects have colonized almost all regions of the earth, from the poles to the tropics and from high mountaintops to deep caverns. Only the oceans contain relatively few insects.

Approximately half of all insect species eat plants. Most of these herbivorous insects are choosy, eating only one or a very few kinds of plants. Moreover, they often eat only the pollen, nectar, leaves, stems, or roots. Every plant species, however, is eaten by at least one or more insect species that have become specialized in consuming that plant.

The other half of the insect species eat other insects, or refuse, or mammal blood. There is probably nothing in the world that cannot be eaten by some kind of insect. Even dead material, such as hair, is food for some: clothes moths readily devour animal fur.

Essential to the Earth

Insects are indispensable inhabitants of the earth because of their enormous diversity and the variety in their life histories. Two-thirds of all flowering plants depend on insects for pollination, and thus for their reproduction. Insects that eat other insects, such as ladybugs, wasps, and ants, can prevent certain insect pests. Insects that clean up refuse, such as dung beetles and carrion flies, ensure that organic material is recycled and doesn't accumulate. And, of course, all these insects are food for other animals, and are therefore important links in the food chain. Songbirds, such as the great tit, depend on insects to feed their young. The songbirds, in turn, are eaten by larger birds, which are then consumed by even larger birds of prey. In short, insects fulfill a number of crucial roles in nature. We could even say that, without insects, there would be no life on Earth. Were humans to die out, insects would live happily on; but without insects, the planet would become uninhabitable: many plants would become extinct, dead animals and dung would pile up, and a great many food chains would be broken.

Humans make use of insects in many ways. Honey bees pollinate our crops and have provided us with a natural sweetener for many centuries; honey is still a much-loved food. We use ladybugs and parasitic wasps for biological control of insect pests in agriculture. Fly maggots are increasingly used in medical care, just as they were in centuries past, for cleaning out and healing wounds. A certain scale insect, the cochineal, is grown and harvested for its bright red color, carmine. This dye is added to give such food products as surimi, red candy, cookies, and strawberry yogurt their color.

Eating Insects

Insects are particularly important as a protein source. If birds, mice, anteaters, and apes eat insects, why should we not eat them as well?

This is actually not a valid question, as people in Asia, Africa, and Latin America do commonly eat insects—not because of hunger, but because they are considered special treats: in the tropics, pound for pound, insects are often more expensive at the market than meat is. This is exactly the same as with shrimp in Europe. In Mexico, people munch on popped ants at the movies. In Singapore, they eat giant water bugs. In China, silk moth pupae and honey bee larvae are regular menu items. Many different insect species are perfect, nutritious ingredients for preparing delicious meals. Over 1,900 species of insects are eaten worldwide! This is the subject of our cookbook: why and how to eat insects, including practical information, recipes, and tips.

Eating Insects: "A Question of Education"

For ten years (1997–2006), he served as secretary-general of the United Nations; Kofi Annan has dined all over the world. Yet, despite the fact that insects are well-appreciated food in many countries, he has never been served a meal that was proudly advertised as including them. Nonetheless, Annan expects that the day is nearing when an esteemed guest will openly be offered an insect delicacy.

Kofi Annan. (Courtesy of the Kofi Annan Foundation)

The goals of the Kofi Annan Foundation include the promotion of sustainable development and food and nutrition security. Kofi Annan has had ample experience with the vast diversity of politicians around the world, and has discussed many urgent political issues with these leaders. During his travels to every corner of the globe, he has partaken of many special dishes. Surely, insects have been included from time to time: insects make up part of the diet of 2 billion people. Annan shares his experience: "When you travel around the world, you do not ask what people are feeding you; you join them as a gesture of friendship. I am sure that during such occasions I have eaten insects. Once, I tasted locusts, which were rather crunchy. Furthermore, I noticed that when locusts swarm the fields, farmers are worried that the insects will destroy their crops, while others are very happy that they will get a valuable source of protein. I think we saw this not long ago in the Middle East and in Israel, where Orthodox Jews were in the fields collecting insects and going back to biblical times."

EDUCATION

Perceptions of insect consumption vary, in Africa as well as in other parts of the world. Some consider it a primitive habit or a poor man's diet. Kofi Annan: "In some parts of Africa they are open to it. Others are not used to it and are not going to buy it at the market. The farmers, the people in the countryside, do eat insects, but those living in the cities do not. They tend to shy away from it."

In the Netherlands, people are getting used to the idea that you can eat insects, and the topic is taken seriously. Annan responds to this news enthusiastically: "I am happy to hear that. It is a question of education. If, today, you raise the possibility of eating insects in the Western world, the reaction is often one of disgust, except among people who have traveled and been around. Yet when you look at the facts—for example, that the human population is growing rapidly, with the expansion of the middle class and the pressures on the traditional protein sources, beef and poultry meat—we will not be able to sustain the demand. Particularly when you consider the imprint of cattle and other livestock on the environment, you are better off with insects. Insects have a very good conversion rate from feed to meat. There is no way that we can sustain conventional livestock production environmentally if we want to meet the needs of the growing human population."

Food provision for 9 billion people is a challenge, especially in terms of animal proteins. "We have to acknowledge that many people—one in eight—do not get enough nutrition in terms of animal proteins, even though they have enough calories to eat. If we can raise insects as an animal protein source, we should be able to bridge that gap. I am sure that we are not only talking about food security, but about food and nutrition security. I believe that if people start to understand it that way, their attitude will change. And the economics will also make people aware, because we cannot continue the way we are producing food: we cannot provide nine billion people with a nutritiously balanced meal in the conventional way."

SHRIMP AND LOCUSTS

Although Annan believes that insects as a source of animal proteins can help us to bridge the gap in nutritional needs that we see today, this will not be achieved without education. That is why he considers the publishing of *The Insect Cookbook* a very good approach. In fact, he tested the book's effect on a member of his staff at the Kofi Annan Foundation. "I said, 'Do you eat insects, or would you eat insects?' I got a recoiled reaction. Then I showed him a picture of one of the dishes in [the Dutch edition of] *The Insect Cookbook* and I asked what the difference was between that dish and a plate of shrimp. The response was, 'Now that you mention it, they are quite alike.'"

In May 2013, the Food and Agriculture Organization of the United Nations (FAO) published the downloadable book *Edible Insects: Future Prospects for Food and Feed Security*. Allegedly, within just twenty-four hours after it was launched, there were already 2.3 million downloads—an indication that education is already taking off, which Kofi Annan finds exciting. "I tell you that we need education, and if you get that kind of response, it demonstrates that there is a genuine interest in edible insects. Also, people who tend to be vegetarian and are staying away from the big mammals because of their imprint on the environment will see the advantages of using insects, rather than cattle, pigs, or poultry, as a valuable source of protein."

REARING INSECTS

Annan asks, "Which countries are best in rearing insects?" This is probably Thailand, because of its 20,000 small domestic cricket farms. "I traveled, as a young man, in Thailand, eating lots of things, and you never know exactly everything you eat. You just say, 'It tastes good.'" At present, there is clearly still some embarrassment surrounding edible insects, but Annan is confident that, in the future, people will be more open in serving these foods. "It will come, I am sure the day will come, and in fact perhaps faster than we think, because we are talking of education: educating the rest of the world, who are not already using insects as nutrition. This is definitely the way to go. I imagine that, because of international travel, within five to ten years insects will be served openly. At an official meeting, one will say, 'Secretary-General (or Prime Minister), this is a delicacy from our part of the world. I know you are going to love it; please taste it and let me know; we have more for you.' Yes, this will definitely come."

FROGS' LEGS

If Westerners start eating insects, aren't people in the tropics likely to say, "Well, we've been doing this for ages?" Kofi Annan: "Yes, that is true. We also need to assure people that there are standards, that it has been scientifically proven to be good for their health and that all the safeguards are met. I recall a moment, years ago, when I was with a young African friend. He happened to be in Geneva, and we had an event where frogs' legs were served. I knew that if I told him what he was going to eat, he would not touch them. We were served when everybody was seated, and he ate them. He liked them: he found them very tender, and well spiced. And then one person said, 'These are frogs' legs!' But because he had eaten them and liked them, he had been initiated—even though, if you would have told him up front, he would have abstained. It is the same with insects. It means that you have to encourage good chefs, and that is why I am very happy to see that chefs are involved now."

What political challenges must be met to incite enthusiasm about eating insects? "For Western people, one has to link it with the environmental advantages that edible insects bring. Link it with the big footprint of cattle rearing. Moreover, you need to get them to understand that it does improve health and is good for development, and also, in a way, is also more economical. Apart from people getting to accept it, you also need to have methods of producing it in commercial quantities, and storing it and shipping it to Europe and America. In Asia, they are used to it. With 20,000 small farms, like in Thailand, you can do it."

INVOLVEMENT OF POLITICIANS

Annan is clear about the path before us. "To move forward, you need to get the politicians on board. They are contacted by the people about the most important issues in their community. Food security is quite important, but in Western countries, most people do not feel that there is a food crisis. Yet there are 2 billion people in other parts of the world who either do not get enough food to eat or lack sufficient nutrition in terms of animal proteins. Still, those who have enough to eat and do have sufficient animal proteins are concerned about the environment, and want to limit the footprint of their food production. For this latter community, I would say that the environment as well as the nutrition are selling points. However, in places where people lack sufficient animal proteins, I would put the nutrition much higher on the list. So it is a question of sensitizing the issues that are important locally. You have a message to sell, and different communities have different needs. In any case, I think you need the politicians, because you have to let them know that this is a natural and valuable source of food that can help to solve important problems."

How, then, to get politicians on board? "I think you have to get the politicians to look to the future, at the demographics. You have to get them to understand the rapid expansion of the middle class. The middle class is expanding very rapidly in the developing world, and you may have millions of Chinese in that category. That will put pressure on the availability of meat and poultry. As a consequence, prices will go up." Annan stresses that we need to emphasize the nutritional aspects of food security and focus on the production of animal proteins without destroying the environment. "You will need to invite politicians to dinner and let them tell the world how delicious it is. If they like it upon their first experience, they will become your best ambassadors. They will proudly go around and say, 'I ate crickets, I ate locusts, and they were delicious.' They will use this to show how courageous and adventurous, and what leaders they are." Does he think that a politician would be more willing to eat insects if told that the former secretary-general of the United Nations is in favor? Kofi Annan: "Let me put it this way, he or she may be shy or unwilling to be proven a coward. Tell them I wish them, 'Bon appétit!'"

Annan is convinced that we need ambassadors for entomophagy, and that we need a mix of people because we are trying to reach a variety of groups. We need individuals who would normally not be associated with insects.

CRICKETS FROM THAILAND

Although insects have been commonly eaten in many parts of the world, the most recent developments in the promotion of edible insects have come from the Netherlands. Annan is not surprised. "The Dutch are so good with agriculture; they have done it for centuries. I am sure that, when they get into this area, the Dutch will probably develop techniques to produce them, and that will go very fast. It is fascinating that European entrepreneurs are busy with it." Still, this is counterintuitive. Why should the Dutch start this enterprise? Would it not be logical for it to develop in those areas where eating insects is common? "The problem is the trade laws. And that is where we need some equity, because sometimes the conditions are so restricted that import from that part of the world is not allowed. I hope that that will change; it really has to change." Should the West promote entomophagy when it is a given in large parts of the world (Africa, Asia, and Latin America)? Annan is convinced that education about human health and nutrition will result in expansion of the market. Eventually, countries will start to import because Europe cannot produce enough. The Thai, for example, have so many farmers that they could crank up their production if given export rights. He considers it important that Europe embrace entomophagy, as he is convinced that this will result in the markets opening up. It is like beef, which you get from Argentina. The next step is crickets from Thailand.

Insects at a market in Chiang Mai, Thailand. (Arnold van Huis)

Cooking with Edible Insects

Henk van Gurp. (Lotte Stekelenburg)

At first, Henk van Gurp, one of the authors of this cookbook, thought that cooking with insects was an exciting challenge. By now, however, it is no trouble at all for him to create delicious dishes and snacks that are enjoyed by many people.

The authors of this book first got together in the 1990s to try to stir up Dutch public interest in eating insects. The Laboratory of Entomology at Wageningen University, in the Netherlands, where Arnold van Huis and Marcel Dicke work, was located very close to Rijn IJssel Vakschool, the hotel and tourism school in Wageningen, where Henk van Gurp teaches. This initial meeting turned into a long-term, unique collaboration. In the last ten years, the partnership has intensified, as interest in the subject has gone into overdrive. The authors worked together in 2006, during the Wageningen—City of Insects festival; in 2008, at the Horecava (an annual national fair for the food industry in Amsterdam); and in 2011, at the Insect Experience Festival, also held in Wageningen. These large events led to enormous media interest within the country as well as beyond its borders. Moreover, the government, other universities, hotel and catering trade press and entrepreneurs, educators, professional and amateur cooks, the food industry, and consumers have all become interested as well.

This cookbook includes more than thirty recipes created by Henk van Gurp and inspired by his interactions and experiences with fellow cooks, colleagues, friends, consumers, and students.

Edible Insects

Several species reared for human consumption are available. This book uses three species sold in the Netherlands:

- Migratory locusts (specific type of grasshoppers)
- Yellow mealworms (beetle larvae)
- Lesser yellow mealworms (beetle larvae), also called buffalo worms, which resemble the yellow mealworm but are smaller

For additional information and suppliers, see page 179.

Tips on Using Insects for Cooking

Arnold van Huis eating crickets in Laos. (Arnold van Huis)

- In the recipes, the word *grasshopper* includes locusts as well as other grasshoppers. The recipes can, of course, be adapted to include whatever food-grade insects are available.
- Most of the insects used for the recipes in this book are freeze-dried, mainly because these are easy to store and easy to use, and they taste good as well.
- Freeze-drying is a safe method of conservation that preserves the insects' color, structure, taste, and nutritional value. The insects regain their original shape when they are rehydrated—for example, by marinating before cooking or adding to sauces. An alternative would be to soak them in warm water or bouillon for about 10 minutes before use. In certain recipes—if roasted, for example—they are used dried.
- Packaged, freeze-dried insects can be kept for long periods, as long as they are stored in a cool, dark place. Check the expiration date on the package. Once the package has been opened, the contents should be kept dry.
- When using grasshoppers, remove the legs and wings first, as these parts are sharp and tough.
- If using fresh (live) insects, it may be wise to mention to the supplier that they are to be used for human consumption. Mealworms and buffalo worms should be rinsed and briefly blanched before use.
- Live grasshoppers should first be frozen and then the legs and wings removed before returning them to room temperature.
- Live insects can be kept in the refrigerator for several days.

- Blanched mealworms and buffalo worms store very well in the freezer.
- Fresh insects have a more subtle and delicate flavor than do freeze-dried ones.
- The most common insect cooking methods are roasting, baking, and frying; these bring out the insects' flavor.
- Dishes prepared with insects taste best when consumed right away. In terms of perishability, rules for storage of food prepared with insects are the same as for any other foods.
- Insects are a good source of protein, so they can be used as a replacement for meat, fish, or chicken. They are great in sweet dishes, with their nutty flavor and extra protein; they also make a tasty and attractive garnish, with their interesting, beautiful shapes.
- You can, of course, adapt or create your own recipes to include insects.
- The amount of insects in any one dish can be varied to taste.
- Insects can be used in dishes both visibly and invisibly.
- Try serving a surprising insect dish or snack at a party; you will be the talk of the town! However, be sure to tell your guests what you are serving.
- Children are often easygoing and quite willing to taste insects. Try tossing a handful of buffalo worms into your batter the next time you make pancakes.

Have fun cooking with and eating insects: use your imagination to create outrageous, tempting meals.

Culinary insects on the shelf in Jan Ruig's poultry shop.
(Lotte Stekelenburg)

"You Have to Eat Away the Fear"

"Absolutely fantastic!" Pierre Wind's hyperenthusiasm is unbeatable. "I'm a true believer," he says about insects. "Think big."

Location: the studio of the children's television program Z@ppLive. That "crazy chef, Pierre Wind" has arrived to cook, a weekly ritual. Once the kids have taken the "I don't like that" oath in which they promise to try whatever Pierre serves them, they embark on a food adventure culminating in mealworm lollipops. Every child in the studio is eagerly licking and munching. "'Icky doesn't exist'; that's my motto. You know what I mean?" Maybe Pierre Wind isn't that crazy after all.

"From the moment the movement started, I've been working with insects," the chef recalls. "At first out of cook's curiosity, but later also because I really started to believe in its value." His introduction to eating insects came in about 1997, thanks to an English cookbook. "That was about playful, funny 'cookology' things. There is strange stuff in there, like cooking with placenta. Because I want to be on top of everything, I was very impressed." Wind began to experiment with insects. "Back then I still had to get my mealworms and grasshoppers from the pet store. Now you can also buy them wholesale."

SENSIBLE

"When the Wageningen professors took this on, eating insects became serious business. The university took it out of the realm of the absurd. I adopted that vision, and all over the country, at schools and in TV programs, I share it. That's because I truly believe that insects are healthy and that we will all be eating them in ten or fifteen years. And if you're sensible, you understand that we will need meat substitutes in the future."

According to the chef, insects have long ceased to be a silly gimmick. He takes them absolutely seriously. "But eating is emotion. We have to try to change the emotion around eating insects, and make it clear that you can just eat them, that they are healthy and delicious. Through marketing. And because eating insects is good for the environment, government should play a key role in putting insects on the market." This chef from The Hague remembers in detail that former minister Gerda

Verburg made it possible for insects to be served in the cafeteria of her department, the Dutch Ministry of Agriculture, Nature, and Food Quality. "A terrific move."

ZOO

"At first it may feel eccentric, like, 'Yikes, what is this?', but once you've heard the story behind insect consumption, you just have to take it seriously." One example: "If you see zebra on a menu, it makes you think of the zoo. But it is eaten in so many countries. You know what I mean? Our culture has to change, and that will take years. People think insects are creepy or bad for you; they know nothing about insects, so if you serve some up, they think, 'Get lost, we're not eating that.' Their body reacts; they feel resistance. That's why you have to tell the story behind the insects." And suddenly grasshoppers and mealworms aren't so strange after all. "Look, if you buy a cow from the butcher, it's not alive. But if you were to see the animal grazing in a pasture, you wouldn't want it. The same is true for mealworms. You can get them wholesale, freeze-dried. Nothing disgusting about them, but you probably have a whole different image in your mind.

"The second step is for people to try them and like them. It has to be really delicious, because if people taste something for the first time and don't like it, they never will. That's why marketing is so important: how are you going to present it? Get a few smart minds together, make an insect introduction plan, and implement it."

FOOD DESIGN

Pierre Wind, who teaches food design at HAS University of Applied Sciences in Den Bosch, the Netherlands, explains: "There are two types of design. With one, you start with the idea and then look at what you can do with it, or whether there is a market for it. With the other, you start by determining who your target audience is and what you want to achieve. That is a more technical approach.

"Every group in our society sees products differently. You can make something for the nouveau riche or make something really inexpensive. Or something truly delicious for the finest restaurants. These are all different approaches. You can also make a cookie or a Thai cake with mealworms, something just for supermarkets. The *mariakaakje* [a plain Dutch cookie] is the most ideal product in the world of food design. It's very simple, timeless, and has a large target audience. The food designer has to decide on two more things: Should the insect product be used for its nutritional value, or as an emotional experience? In other words: should it be incorporated visibly or invisibly? You can think up the greatest things. If you want to capture the market with insects, then you have to get a beautiful product on the menus of top restaurants, and also reach the public at large." He knows that Lollo Rosso lettuce became a success that way. "Think big."

THE WHEEL

"The nice thing about this topic is that you don't have to prove anything. We don't have to reinvent the wheel, because insects are eaten in many countries already. I ordered nice, crispy cookies from Thailand baked with insects that you could see if you held a cookie up to the light. It's a delicacy over there." Another comparison: tofu. "It is produced because we need meat substitutes, but you might think, 'What a disgusting soy thing.' You have to learn what to do with it; you have to *make* it taste good. Maybe someone once thought, seeing the very first slab of bacon, 'What the heck is this?' You know what I mean?"

Boy eating mealworm ice cream at the Insect Experience Festival in Wageningen, the Netherlands, 2011. (Rudolf Barkhuijsen)

On television, previously on *The Food Factory* (*De Eetfabriek*) and now on *Z@ppLive* and other programs, you frequently see Pierre Wind using insects. "Children are attracted to them in droves. They are more open-minded and adventurous than adults. That's why you have to start as early as possible with taste lessons." Insects, he believes, should be included. "It's just a matter of education. First you have to tell people that insects are edible and healthy; only then does it become interesting that they also taste good. You have to eat away the fear."

FUNCTIONALITY

"Nutritional value is an important reason for eating insects. The taste experience centers primarily around their crispiness. That provides a nice 'cookological' effect in cookies, for example. And you can garnish with them. That looks beautiful." But the point is that they're functional. "Just like cornstarch: you use it only for its functionality. As a binder, not for its taste."

Grasshoppers, mealworms, or buffalo worms: to Pierre Wind, they're all the same "The preparation is the same. You have to sauté them first and then continue with the preparation, in your quiche, mixed into the fried rice, or as a garnish on a pie. Also delicious: deep-fried and sprinkled with powdered sugar. Or—mealworms are especially good for this—you can puree them as a base for a filling of some kind. Grasshoppers can be used that way, too, but if you were to ask me which insect is more user-friendly for 'cookology,' I would choose the mealworm. Because grasshoppers contain a little bit of air. If you fry them and allow them to stand, they become tough." And no, the grasshopper is not pretty. "Get off it. Yeah, maybe for nature lovers. But they should go see them in nature. The mealworm, with all those rings on its body, is much prettier." Then, more calmly: "I look at it 'cookologically,' right? You can't eat the legs, so you won't see them in the dish. Unless you fry them whole and nibble them off, but I think you shouldn't go that route.

My crystal cooking ball predicts that this will be a success. In ten or fifteen years, insects will be sold in supermarkets. I see a big future for insects."

Children eating insect snacks in Laos. (Thomas Calame)

Everyone Eats Insects

Insects are everywhere. They are found in all our food crops, despite the fact that farmers and growers do their best to protect them. At harvest time, the best apples are sent to markets and shops, and lesser-quality apples are used to produce apple juice and applesauce. Apples sometimes have an insect or two in them—and these just get ground up along with the apples and become part of the applesauce and juice. The same goes for tomatoes and ketchup, grains and bread, coffee beans and coffee, and a long list of other foods. Inspection authorities do enforce specific limits. In the United States, for example, the maximum is sixty insect pieces per 3.5 ounces (100 g) of chocolate, thirty insect pieces per 3.5 ounces of peanut butter, and five fruit flies per 1 cup (250 ml) of fruit juice. Calculations indicate that each of us unknowingly consumes about 1 pound (500 g) of insects per year. This means that our food is enriched with high-quality protein. It also means that even our vegetarian "spaghetti-without-meatballs" contains a tiny bit of (insect) meat.

Caterpillar on broccoli. (Lotte Stekelenburg)

Shrimp or Grasshopper?

In the Western world, we have learned to eat many new things in the last few decades. For example, in northern Europe, cultured *champignon* mushrooms were practically unknown before the 1960s; southern European vegetables, such as peppers and eggplant, became available only in the 1970s; and kiwis first appeared in the 1980s, whereas sushi is an even more recent import. It is clear that northern Europeans are no longer as attached to the standard diet of meat and potatoes as they once were. They eat mussels, oysters, snails, lobsters, and shrimps as delicacies, and pay good money for them. And eating this seafood is completely normal. But insects on their plate? No way! Take a closer look at a shrimp, however: surprisingly, it looks a lot like a grasshopper! A shrimp has ten legs; a grasshopper, six. A grasshopper has wings; a shrimp has none. But if you take away the legs and wings, the resemblance is striking. Shrimps are a delicacy in Europe and the United

Grasshoppers at a market in Tahoua, Niger. (Annie Monard FAO)

States, while grasshoppers are a delicacy in Africa. In Australia, clever entrepreneurs have even managed to cash in on grasshopper plagues and turn these insects into "sky prawns." The grasshopper has what it takes to conquer the West. If we could overcome our prejudice and introduce the "sky prawn" in our countries, grasshoppers would have a bright future on our plates.

"I Could Eat Insects Anytime, Day or Night"

"They're pretty good, huh?" Her hand disappears again into the container of mealworm cheese sticks. "And the recipe is super simple." Insects are a welcome mouthful to Harmke Klunder, both literally and figuratively. This researcher talks about idealism, termite porridge, and cooking contests.

"Actually, I had a date with a girlfriend yesterday, but I had to give this recipe a try. So I just combined the two, and we ducked into the kitchen together." That happens quite often. Harmke Klunder's friends have now gotten used to being served insects once in a while. "By now, most of them have no trouble popping a mealworm or grasshopper into their mouth. Some even try it out at home, and e-mail me to ask if I can come up with a recipe for their birthday party or something like that. But there are also people who eat insects just to humor me."

It does, in fact, make her happy to help others discover insects. "Mostly out of idealism." That became evident to her while studying food technology. "It was always about apple juice or orange juice. But when I looked in the supermarket, I wondered, 'Is a new package of apple juice really necessary?'" Klunder decided to interrupt her studies for a year of travel. "Then I thought: 'Something in development work?'" That wasn't it, either. She went back to her studies. An instructor asked her to develop the new course called Food Cultures and Customs. "While I was working on that, I met a Ph.D. student who was studying insect cells as a source of edible protein." Interesting, she thought. "But then that information about edible insects disappeared somewhere in an inactive part of my brain."

GLOBAL HUNGER PROBLEM

Only later—during her specialization in food security—did that part of her brain become activated, she says with a laugh: thanks to a contest to devise a food-technology approach to the world's hunger problem. "Just perfect for me." She went to work on it with fellow students from Mexico, Poland, and Turkey. "One of them said, 'We should do something in Africa.' Another: 'We should use sorghum [a grain] instead of corn.' The third: 'But it has to be available locally.' And I thought that we should do something with insects. Because I knew that termites are a common food in Africa, the insect species was obvious. Immediately I felt, 'We've got something here.'"

RESEARCH

The idea of a fermented porridge of sorghum and ground termites for added protein proved simple and good enough to be allowed into the contest's final round in the United States. Eventually, they even won first prize, with just an idea on paper. Strangely enough, Klunder didn't actually make the porridge until after the contest, while doing a senior research project on the effects of adding insects to fermented foods. Finally she tasted her own sorghum-termite porridge. "It lived up to my expectations: very sour." Naturally, she hoped that the idea would actually be introduced in Africa in the future. "But to turn it into a commercial product, you need a fortune."

During her research, Klunder saw mealworms for the first time. "The termites had to be imported from Africa, but because I had to start my experiments, I began with mealworms from the producer Van de Ven. They just gave me a few pounds of the live animals. Yeah, and then I had those mealworms sitting on my table. Funny. I thought, 'I have them; so, now what?' I started to experiment in the kitchen: boiling, frying, blanching, steaming, but I felt so sorry for them. Then I put them in the refrigerator before preparing them, to make them sluggish. But once out, they came to life again, so it was still pitiful to heat them. You're killing so many souls, as it were. I actually thought that I could hear them screaming in the pan. Finally I froze them, because I think that's the nicest way to die. First they get sleepy, and they don't notice that they are dying."

FAO

The stage was set: the former student would be working with insects for the foreseeable future. Very cool, she thought. "I could eat insects anytime, day or night." That feeling intensified when she went to Laos for an internship with the FAO (Food and Agriculture Organization of the United Nations). There, she helped set up insect farms following food safety standards. "In Vientiane, the capital, there were little tents everywhere selling insects as snacks. Laotian men often went directly from their work to drink beer and munch on all kinds of insects. I did that, too, of course. I tasted everything there. Laotians don't have any barriers at all about eating insects. Then I noticed this crazy thing: sure, I was very much in favor of them in theory, but in practice there was still something holding me back. A Laotian friend and I cooked in the evenings, using the insects left over from the laboratory. The kinds of insects were always changing, because they are seasonal. You know what's delicious? Wasp larvae. You can steam them, then add salt and pepper. The taste is somewhere between a bean and a nut. And weaver ant pupae. Those look kind of like coarse white caviar. I liked to put them in the *tom yum* soup, instead of shrimp."

In Laos, she created a recipe for insect noodles. "I added insects to a very popular dish. Simply a little bit less noodle dough and a bit more ground insects. That way, you take in more protein. In those countries, protein deficiency is a real problem. Those who eat too little protein get sick."

COOKING CONTEST

During Klunder's internship, the FAO organized an insect-cooking contest for restaurants. "That was very special, as I had been interested in this topic for such a long time. On the day of the contest, cooks from fifteen restaurants suddenly appeared, all of them using insects— whether cooking Laotian, Western, or Japanese style." The winning dish was garlic-scented crickets in a crunchy taco shell with silkworm cocoons, sour cream, and spicy tomato sauce. "Those cocoons have a fairly high percentage of fat, so the sauce is extra creamy."

QUALITY ENGINEER

Klunder's travels eventually took her via America, Laos, and Wageningen to Oostzaan, north of Amsterdam, where she is now employed as a quality engineer by Jan Ruig's poultry company. Her job involves food safety and product development. "Ruig makes special products,

Winner of an insect-cooking contest. (Thomas Calame)

but until now all of them have been meat-based. An insect is different. That's why we are now looking for a product without meat but with protein, perhaps with legumes or something similar. You know, this is really exciting, because all of the experimenting I've done until now was all well and good, but a product has to make money for a company; it has to satisfy the demands of the market. There is barely any market for insects as yet, so we don't know what those demands are." In any case, she is striving to develop a product that is as sustainable and environmentally sound as possible. "Maybe something falafel-like? Or something very different." The suspense continues.

Bugsy Cheese Sticks. (Lotte Stekelenburg)

This researcher estimates that it could take another ten years for the Netherlands to be fully "insect-ready." "But my own perception, and that of my friends, has already changed, so it must be possible for others as well. We all really will be eating insects. As long as they taste good." Her eyes sparkling, she looks down at the container of cheese sticks. It's now empty.

Bugsy Cheese Sticks

Equipment needed: baking sheet lined with parchment paper

4 (5-inch [13 cm]) squares frozen puff pastry dough
¼ cup (10 g) buffalo worms
1 egg, lightly beaten
1 tablespoon sesame seeds
Sea salt and freshly ground black pepper
5 tablespoons grated aged cheese

Preheat the oven to 350°F (180°C).

- Make a rectangle by overlapping the edges of two squares of puff pastry. Moisten the edges with a little water and press together. Make a second rectangle with the other two squares.

- Brush the rectangles with egg and sprinkle them with the buffalo worms, sesame seeds, sea salt, and pepper. Finally, sprinkle with cheese and roll the top lightly with a rolling pin to make sure all the toppings stick well.
- Slice each rectangle into five long strips, and twist each strip three times to create a spiral effect.
- Transfer the strips to the prepared baking sheet and bake in the middle of the oven for 15 minutes, until golden brown. Allow to cool, then enjoy!

YIELD: 10 STICKS

Public tasting insect delicacies in Laos, 2010.
(Thomas Calame)

Weaver Ants in Asia

Weaver ants are ants that build their nests in trees. They pull leaves together, attaching (weaving) them together with silk threads. Larvae and pupae of these weaver ants (of the genus *Oecophylla*) are a very popular food in Asia. They are sometimes called ant eggs, as the pupae resemble eggs. People prefer to eat the larger larvae and pupae; these are the future ant queens that—if they escape being eaten—fly out to form new ant colonies.

Weaver ant nest. (Arnold van Huis)

Weaver ant larvae and pupae can be found in the nests during a few months of the year. They are usually harvested by women, who wear protective clothing and gloves, as the ants can bite viciously. In Laos, women go out in groups to look for trees containing nests. They carry long bamboo poles to which they have attached a bag or a basket. Using the poles, the women poke at the ant nests, which are often found high in the trees. The contents of the nest fall into the container, and are then shaken onto a platter and sorted by blowing away the adult ants, leaving larvae and pupae behind. These are either used for the women's own consumption, or sold to market women, earning the collectors a pretty penny. At the market, the weaver ants are usually placed on ice to keep them fresh. People like to add them to soups, salads, or fried dishes. Well seasoned, they also make a nice raw snack.

Women setting off to gather ant nests. (Arnold van Huis)

Wasp Larvae in Japan

To us, wasps are nasty stinging beasts that annoy us when we sit outside having a drink in late summer. These are usually the wasps known as yellow jackets. In Japan, however, the same wasps are considered to be

Weaver ant pupae at a market in Laos. (Arnold van Huis)

exquisite treats. Emperor Hirohito's favorite dish was rice with wasp larvae cooked in sugar and soy sauce.

Wasp larvae are eaten mostly in the central mountainous areas of Japan. The nests are found on the forested slopes. Men have made a sport of collecting wasps, and have thought up an ingenious method for it. Wasps are carnivorous and eat many other kinds of insects, and are also not averse to eating a piece of meat or fish. The wasp hunters attract the wasps with a small piece of meat to which they have attached a tiny white flag. As the wasp grabs the meat and flies away, the men can follow the little white flag, running, climbing, and scrambling through the woods in pursuit of the wasp. After locating the nest, the men use smoke to anesthetize the insects and quickly dig out the nest. The cells within the nest are filled with larvae, ready for consumption.

A piece of meat with a white flag is offered to a wasp. (Kenichi Nonaka)

The wasp hunters usually leave a portion of the nest intact, to ensure the colony's survival.

Many Japanese people grow wasps in their own backyards. Nests are put in a sheltered place, and the wasps are fed meat, fish, and sugar water. In the fall, when the nests are fully grown, the larvae and pupae are harvested for food. There are many recipes, but most people just cook them with soy sauce, adding sugar and sake (rice wine), and eat them with cooked rice. Canned wasp larvae and pupae can be found in Japanese supermarkets, and they are considered a costly delicacy.

Canned wasp larvae in a Japanese supermarket. (Kenichi Nonaka)

Every year, a wasp festival takes place in the central highlands of Japan, during which there are wasp cooking competitions and the sale of wasp products. A prize is also awarded to the person having collected or grown the heaviest wasp nest.

Termites: A Royal Meal

Termites are social insects that live in a nest; their colony consists of a king and queen and many soldiers and workers. Although an individual termite may weigh about 2 milligrams, the combined weight of all the termites in the world exceeds that of all the humans put together.

Termites live predominantly in the tropical and subtropical areas of Africa, Asia, the Americas, and Australia. They are known for their large nest hills, which can reach a height of 25 feet (8 m). Termites are considered pests, as they excavate wood from the inside, destroying buildings. Australians call termites white ants, but ants and termites belong to different insect groups, even though they are both social insects.

In the tropics, termites are considered to be a special and nutritious food. They contain a great deal of protein, essential amino and fatty acids, iron, calcium, and other micronutrients. Termites are often fried in their own fat; the leftover oil can even be reused to prepare other meat dishes. Sometimes, termites are wrapped in banana leaves and steamed. After cooking, they can be dried for conservation. In Kenya, people make cakes using powdered, sun-dried termites to replace up to 5 percent of the wheat flour.

Generally, the termites that are eaten are the future kings and queens. At the end of the dry season and after the first soaking rain of the rainy season, the winged ones leave the nest in large swarms.

Males and females mate high in the air, and then fall to the ground and pair up to start a new colony. In Kenya, people try to trick the termites into believing the first clattering rains have come by drumming with sticks on the ground surrounding the nests, which makes the termites appear. When the termites land after their mating flight, they immediately lose their wings, which makes them easier to catch. A bowl of water is sometimes placed under a lamp, attracting flying termites that then fall into the water, lose their wings, and can be simply scooped up by the handful. People either dry them for storage, or fry them right away—no oil needed, as the insects are fatty enough.

Termite queen. (U.S. Department of Agriculture)

Soldier termites are also eaten. They have huge jaws, and their heads make up about half of their entire body size. In Uganda, children are adept at catching termites. After chipping off a piece of a termite hill, children extract the insects by lowering grass blades into the shafts of the opened termite mound. In defense, the soldier termites bite into the blades and are subsequently stripped from the blades into a container. The termites caught can be fried or pounded into a cake, and added to maize porridge or cassava porridge.

A termite queen produces about 150 eggs each minute. Her abdomen is 4 inches (10 cm) long and is full of eggs. Because of their high concentration of valuable nutrients, termite queens are a very healthy food and are often given to undernourished children. Removing the queen from a termite hill means the death of the colony, however, as the queen is the only one that can produce offspring.

Termites after their mating flight. (Christiaan Kooyman)

Termite hill in Uganda. (Arnold van Huis)

Termite Snack

½ cup (20 g) wingless termites
Pinch of salt
2 pinches of freshly ground black pepper
¼ teaspoon curry powder

- Sprinkle the spices onto the termites and fry them in their own body fat in a large skillet over low heat.
- Stir gently to cook evenly.
- They are ready to eat when they are lightly browned and crispy.

YIELD: ½ CUP

Lake Flies in East Africa

Many types of flies depend on water for their reproduction. The larvae of lake flies live in the water of the large East African lakes, including Lake Tanganyika, Lake Malawi, and Lake Victoria. At new moon, when the larvae have pupated, the pupae float to the lake surface and the flies emerge simultaneously, by the millions, forming huge clouds over the water. Arnold van Huis, one of the authors of this book, witnessed this fascinating spectacle once, while having breakfast at a hotel at the shore of Lake Malawi. He thought that a fire had broken out, because he saw huge "smoke" plumes rolling over the water. He later found out that these were swarms of lake flies.

Swarm of lake flies in Malawi. (Midori Yajima)

When the fly clouds are observed over the water, the local people prepare to collect them. Large, flat baskets are attached to handles and moistened. As the clouds of flies are blown to the lake-shore, people whirl the basket back and forth through the swarming flies, causing layers of flies to get caked onto the inside of the basket. The baskets are then emptied directly into pans of boiling water. The water is then filtered out, and the fly cakes are dried in the sun. These *kungu* cakes can be kept for some time, and are considered a local delicacy.

Boys carrying a bowl of *kungu* cakes in Malawi. (Midori Yajima)

David Livingstone (1813–1873) described these edible lake flies in an account of his explorations in Africa. The protein content of these fly cakes is very high, as much as 67 percent, making them one of the most protein-rich edible insects. The lake flies are also a good source of minerals.

EDOARDO RAMOS ANAYA, STORYTELLER

"The Tortillas from Way Back When"

Two boys sit on their haunches on the sand under a tree, against the backdrop of the Mayan pyramids of the sun and the moon. A dog barks in the distance. The boys are happily munching on a tortilla with chapulines: *a tortilla with grasshoppers.*

Edoardo Ramos Anaya recalls vividly how, as a boy, he used to picnic near his village, San Juan de Teotihuacán. "When you eat something, the flavor brings you back to the place where you first tasted it. I travel that way every day." He tells this while his hand makes an indentation in a corn tortilla that he then fills with grasshoppers au gratin. The Mexican meal triggers in him both emotions and memories. "When I used to walk through the countryside with my grandfather's sheep on the weekends, I had nothing to do, so I would just collect insects." *Gusanos de maguey*, agave-infesting caterpillars, he could simply pick off the plants; the same was true of *cumiles*, beetlelike creatures with a "fresh Mentos taste." He would catch grasshoppers with a hat or a shopping bag. *Whap*—he demonstrates the motion of using the bag as a net. "At the end of the rainy season, when everything was green, you could see them jumping everywhere. If you were to hang around among the plants all day, you could catch a whole bag full." Occasionally, he would pop a raw one into his mouth. "Just pull off the legs and eat it as is. Fresh grasshoppers are green and soft—a bit like chewing gum." But usually he covered them with a scarf, to keep even a single one from escaping, and brought them to his mother. She prepared them as a meal or made them into a snack. Or she added them to the soup, as croutons.

RESPECT

"In our day-to-day lives, we ate a lot of insects." Self-caught. Ideal, according to Ramos Anaya, "if you don't have much money," but in Mexican culture, food means more than just its economic value. "The connection to the earth is extremely important to us. You take something from it, but you also have to give something back. If you don't do that, the cycle is broken and you die. Everything must be in balance." That is why Mexicans pray for fertile land, why they ask the earth for permission to sow, and why they scatter fried chicken over their land. "Everything that grows through that is edible," Ramos Anaya knows. "It is not a question of spirituality per se, but of respect. It may be difficult for people of other cultures to understand, but the earth is your mother. If you grow corn in Mexico, as we do, then you stay in one place and, generation after generation, you learn everything about snakes, plants, beetles and grasshoppers. It takes a long time before your corn is ripe enough to harvest, so you have to make sure you have something to eat in the meantime." Other plants, such as zucchini. "And insects. They are a threat to the harvest but we also like to eat them." By catching them, then, you kill two birds with one stone. "Just like fungi, which destroy the corn but are themselves edible." The principle is: take good care of your land, and you will always have food.

SANDPAPER

The little critters are especially prominent on the menu in rural areas in the south of Mexico. According to Ramos Anaya, that's the secret to the health of some villagers. "I was in a village near Oaxaca in 1993 where the people had no money for a pig or cow. They ate only tortillas with insects. Those people lived to be very old, to ninety or beyond." That doesn't surprise him in the least. "Meat is junk. If you eat it every day, it stops up your intestines like clay. Insects, just like fiber, are like sandpaper. They clean out your intestines.

"Mexicans who live in the city aren't used to eating insects. They come to Oaxaca as tourists and eat them at one of the little restaurants. For them, it's exotic. There is a very old marketplace in Mexico City that goes back to the pre-Columbian era, where you can sometimes buy grasshoppers if you're very lucky.

Food is culture. Unfortunately, some Mexicans are not proud of their culture. They would rather be like Americans." He grimaces. "They say, 'Grasshoppers, that's for the Indians, not for us.' I, on the other hand, want to show everyone how rich the Mexican culture is." Cooking with insects is part of that. "We have to make it clear to others that insects are not dirty or ugly. Especially children. They are feisty; they have more *power* than their parents. If they learn to eat insects, then the adults will, too. It won't work the other way around."

PORK

He does understand that the Dutch are amazed to see grasshoppers or mealworms on their plates. "The Netherlands has the bad fortune to be located on this little piece of the earth. A long time ago, you ate only turnips, potatoes, and pork. That is your culture. People don't understand that a grasshopper lives in a very clean landscape; they are uneducated and think that insects are dirty. But a pig runs around in its own manure. That is much worse."

Chapulines, or Mexican grasshoppers. (Lotte Stekelenburg)

Shaping a tortilla. (Lotte Stekelenburg)

Filling the tortilla with green guacamole and grasshoppers. (Lotte Stekelenburg)

It came to him as a pleasant surprise that (unlike in Mexico) insects are available wholesale in the Netherlands. "It's becoming hip here," he predicts. "And the crazy thing is, if it's popular in the West, the Mexicans will want it, too. Then it will come back to us via America. Like a boomerang."

He spoons a healthy blob of guacamole onto his tortilla, then some *salsa roja* and two grasshoppers as a garnish. When he takes a bite, he closes his eyes for a moment. The insects he is cooking with here may be bigger and less fresh than the ones his mother used, but they still bring him back to the memories of home.

CHAPULINES AL AJILLO

Ramos Anaya prepares the grasshoppers just as his ancestors have done for centuries. Over high heat, he fries a handful of grasshoppers (without legs and wings) in a bit of oil and lots of chopped garlic. He serves the fried *chapulines* in *casolitas* (small earthenware dishes), for example, as *chapulines queso gratinado* (covered with grated cheese and baked au gratin). His favorite way to eat grasshoppers is with tortilla chips or warm corn tortillas, guacamole, and jalapeños.

Inspired by this authentic Mexican preparation, Henk van Gurp made Mexican Chapulines (see page 43) and filled tortillas, or Bugitos (see page 129).

Spirited Caterpillars in Mexico

Almost everyone has heard of the Latin American hard liquor that features a 2-inch (5 cm) "worm"—actually, a caterpillar—at the bottom of the bottle. Every liquor store has this product on the shelf. The drink is called *mezcal*; it is distilled from 100 percent Mexican agave plants and has an alcohol content of at least 45 percent. *Mezcal* is produced mainly in the state of Oaxaca, where thirty kinds of agave are used to make the various types. It should not be confused with tequila, which is only 51 percent agave and comes without a caterpillar. The caterpillars in *mezcal* are actually an agave pest. They chew on the stems and leaves of the plants, yet they are not controlled with pesticides; rather, they are collected and used in traditional Oaxacan cooking. Not only do they appear at the bottom of *mezcal* bottles, but they are also considered a

Mezcal with caterpillar. (Lotte Stekelenburg)

delicacy: fried and eaten on a tortilla with guacamole, or ground up and combined with tomatoes and chilies in a spicy salsa. The caterpillar in the *mezcal* is a trademark, "*con gusano*," serving as proof that the agave drink is genuine.

Long-Horned Grasshoppers in East Africa

Grasshoppers are such a coveted delicacy in Uganda that they cause traffic accidents. The insect in question is a green long-horned grasshopper, or *nsenene* as it is called in Luganda, the language of the largest Ugandan ethnic group. Grasshoppers are so plentiful in November, and this insect is so popular, that November is known as grasshopper month. Women and children collect the grasshoppers that are attracted to lampposts at night. Collecting grasshoppers along roads with traffic speeding by is perilous and results in accidents every

Gusanos de maguey, or agave caterpillars. (Andy Sadler)

year. Grasshoppers are a good source of income for the collectors; they are worth about thirty cents per pound. At the marketplace, the price is about five times that. In comparison, beef is cheaper than grasshoppers by half.

Professional collectors use a more efficient method. They focus a spotlight backed with a corrugated aluminum funnel up toward the sky. The grasshoppers, attracted to the light, hit the brightly reflective metal and drop down the funnel into a barrel.

Children with strings of grasshoppers in Indonesia. (Annie Monard FAO)

Grasshoppers at a market in Tahoua, Niger. (Annie Monard FAO)

Grasshoppers are also widely collected in the countryside, not only in Uganda but also in Tanzania and other East African countries. When the first grasshoppers are found in the fields, the news spreads like wildfire through the villages. Sometimes, fights break out when outsiders try to catch the grasshoppers in the villagers' fields.

Preparing grasshoppers for consumption is easy: simply remove the insects' legs and wings, and fry. No extra oil is needed because the grasshoppers have plenty of body fat. Enjoy!

"Insects Are Buzzing All Around Me"

An inventive chef at an inventive company, Johan Verbon is the chef de cuisine *at the Restaurant of the Future in Wageningen, the Netherlands. He likes insects—in his kitchen and on his plate. That's the truth right now, and he looks forward to even more in the future.*

"During the twenty-four years that I have worked for Sodexo, the catering firm, I have always made food and product development my business." The meat croquette that you can get at any Dutch company cafeteria, the pizza toppings, the soups: they are all a product of the chef's ingenuity. And yes, they do carry a genuine Johan Verbon "signature": "All my products are sustainable; if possible, organic; and preferably also produced locally." This is true for the famous company cafeteria croquette, made with meat from humanely raised animals and 100 percent organic.

Johan Verbon. (Lotte Stekelenburg)

LABORATORY

It's no surprise that Sodexo management came to see Verbon when they sought a *chef de cuisine* for the Restaurant of the Future. "This isn't a restaurant; it's a laboratory. Everyone can come and eat here, and they automatically become part of our experiments. The Restaurant of the Future collaborates with Wageningen University on consumer research in the broadest sense. One example: if the government says we should eat less salt, we want to know if that is really true, and also how it can be accomplished. What happens, for example, if you leave out 25 percent of the salt in a soup? Will people just add more themselves? It turns out that no one notices if you make bread with 50 percent less salt. But what does it do to the shelf life of the product and how the dough rises? Those kinds of questions are addressed here."

Alternative protein sources are also a focus of attention in the Wageningen food laboratory. "Vegetable proteins, for example, or milk proteins." In which category insects belong is debatable. Verbon: "I suspect that they are in a separate category, alongside vegetable and animal proteins. To me, an insect is an alternative to animal proteins, but a vegetarian might not eat insects. An insect is also an animal. But to me, it's not meat. It is an alternative protein source; that's what I emphasize time and again."

GUT FEELING

"I think that consumers aren't looking to insects for food, but that they would like an alternative for meat. So far, this is based on my gut feeling; I haven't researched it scientifically. I have, however, cooked with insects on a regular basis." You might say that, in his own way, Verbon has tested the new ingredient. "First I searched the Internet and the literature. In such countries as Thailand and Vietnam, of course, insects are eaten in abundance. Whenever I saw a photo of a dish somewhere, I took a good look at it and tried to turn it into a Dutch recipe. Take noodles with grasshoppers, for example."

The public at large had their chance during the 2011 Insect Experience Festival in Wageningen, where a stand offering *poffertjes* (small, fluffy pancakes) with visible mealworms couldn't keep up with the demand. "I discovered that it's easier to get children to eat insects than it is to get adults to, and that parents follow suit if they see their children eat them. And I discovered that you always have to explain that they are insects and that they taste good.

I am one of the few chefs who has experience with insects as a food. Very few colleagues dare. And yes, I am right at the source in Wageningen, of course, with the entomologists around the corner. Insects are buzzing all around me."

PESTO

Not only are *poffertjes* a success, but also "a nice pasta with mealworms," and a pesto using the little mealworms instead of pine nuts. "Or try incorporating them into meatballs. You do need fresh ones, though; otherwise they get tough," Verbon advises. "Still, you're not likely to hear someone say, 'Hey, yummy, a handful of mealworms.'"

He does like the idea of a bag of grasshoppers, though. "I like sweet preparations, such as caramelized or sugared grasshoppers. They're crispy; they have a good texture. Just like popcorn."

What makes something delicious? "To be delicious, it has to be attractive. It has to look good, smell good, feel good, have a nice crunch in the mouth, for example; and it can't have a strong aftertaste. Something like cilantro: either you like it or you don't, because it has such a distinctive taste. Everyone likes the taste of insects. A buffalo worm tastes just like bacon, maybe because you fry both of them. And for a meal, you're always searching for a balance. You need something hard and something soft, something sweet and something sour, and so on." The chef has discovered even more. A few tips: "For mealworms, you should use primarily butter, because that makes them nice and creamy. But I prefer to fry grasshoppers in oil, in nut oil, for example. That can tolerate high heat, which makes extra-delicious crispiness and a nice brown color possible."

HIDDEN

"At first, I thought: 'The Dutch are never going to eat visible insects. No, you have to process them into something, use them undercover.' In my first dishes I had hidden ground insects. In spaghetti flour, for example. And then I served them with a good vegetarian sauce," the chef recalls, "so you took in some protein without really noticing. But I am slowly moving toward the idea that we might be able to present visible insects after all. People eat them more easily than I thought. Until now, everyone I've given insects to eat has said, 'Good, but I did have some resistance to overcome.'"

Verbon would like to do more with insects. "I haven't received this research question, but one thing I would like to know is how we should present them. On what grounds would someone choose potatoes-vegetables-grasshoppers instead of potatoes-vegetables-meatball? At first, probably 99 percent would go for the latter. So then you substitute a certain percentage of insects for the meat, and you give it a label: maybe you sell it as a sustainable product. And finally you do a test in which the communication emphasizes the fact that it contains insects. Communication, after all, is the key to success. You have a choice: Do you sell something without a label, or labeled 'organic' or 'sustainable'? It turns out that people choose what they are used to. Of course, you shouldn't try to fool consumers. No, you have to tell them honestly what is in it, but it's important to tell it in a catchy way. So you don't say, 'This contains 25 percent grasshoppers,' but that it is hip, new, sustainable, that it traveled very few food miles; and then in the small print you mention that it contains insects. The consumer does not want to be educated."

And then, when the stories have been told and the consumer has made a choice? "Then maybe in five years it will be normal for a movie theater to have not only a popcorn stand, but also a grasshopper booth."

Caramelized Grasshoppers

2 tablespoons nut (not olive) oil
About 16 grasshoppers, legs and wings removed
2 tablespoons sugar

Caution: Caramel gets scalding hot, so don't lick the spoon!

- Heat the oil in a skillet over medium-high heat and stir-fry the grasshoppers for a few minutes, until golden brown.
- Add the sugar and continue to stir until the sugar is golden brown as well; this should take about 4 minutes. Be careful, as this step goes very quickly. (Has it turned dark brown? Then it's burned and bitter; discard and start again.) Remove from the heat.
- Allow the caramel to cool and treat yourself and your guests to a deliciously sweet snack. These are wonderful with a cup of tea or served with a sweet dessert.

Caramelized Grasshoppers. (Lotte Stekelenburg)

FIVE SNACKS

Mexican Chapulines. (Floris Scheplitz)

Mexican Chapulines

Equipment needed:
oval ovenproof dish,
approximately 10 inches
(23 cm) in diameter

GUACAMOLE

1 ripe avocado
1 garlic clove, crushed
1 teaspoon Dijon
  mustard
1 tablespoon
  mayonnaise
Juice of ¼ lemon
Pinch of salt
Freshly ground white
  pepper

CHAPULINES

1 tablespoon lime oil
24 grasshoppers, legs
  and wings removed
¼ teaspoon chili
  powder
Pinch of salt
1 5.5 oz (150 g)
  package tortilla chips
1 green chili pepper,
  seeded and diced
2 spring onions, sliced
  into rings
2 tomatoes, seeded
  and diced
¼ cup (70 g) chili sauce
¼ cup (30 g) grated
  aged cheese

¼ cup (60 g) crème
  fraîche or sour cream

Preheat the oven to 325°F (160°C).

- Prepare the guacamole: Cut the avocado in half, remove the pit, peel, and dice. Add the remaining guacamole ingredients to the avocado and puree with a stick blender or in a food processor. Season to taste with salt and white pepper.
- Prepare the *chapulines*: Heat the lime oil in a wok over medium-high heat and fry the grasshoppers for 2 minutes, until crispy. Transfer them to a platter and season with a bit of salt and chili powder.
- Place the tortilla chips in the ovenproof dish. Spread the grasshoppers, chili pepper, spring onions, tomatoes, and half of the cheese over them. Shake the dish to mix, then cover with the rest of the cheese.
- Bake for 15 minutes. Remove from the oven, spread the crème fraîche over the *chapulines*, and serve right away, with the guacamole on the side.

YIELD: SERVES 8

Tips

- The hot green chili pepper can be replaced with mild red peppers.
- The dish can be made spicier by using hotter peppers and/or sprinkling with some Tabasco sauce.

Dim Sum

Equipment needed: deep fryer, with vegetable oil for frying

POUCHES

2 tablespoons sunflower oil
¼ cup (40 g) raw arborio rice
¼ cup coconut milk
½ cup (120 ml) vegetable stock (homemade or prepared from a bouillon cube)
1 shallot, minced
1 tablespoon buffalo worms
1 teaspoon *sambal oelek* (chili paste)
8 wonton wrappers
Small amount of egg white

"CANDY WRAPS"

1 tablespoon nut oil
1 shallot, minced
1 garlic clove, crushed
3½ ounces (100 g) shiitake mushrooms, cut into thin strips
¼ teaspoon curry powder
1 tablespoon soy sauce
8 grasshoppers, legs and wings removed
1 tablespoon sunflower oil
8 wonton wrappers
Small amount of egg white

- Prepare the pouches: Heat 1 tablespoon of the sunflower oil in a medium pan and fry the rice over low heat for 1 minute. Add the coconut milk and half of the stock and cook the rice for 15 minutes, stirring regularly. When the liquid has been absorbed, add the rest of the stock and continue to cook until the rice has absorbed all the liquid and is tender. Remove from the heat, cover the pan, and allow the rice to cool.

- Heat the remaining tablespoon of sunflower oil in a skillet and fry the shallot and buffalo worms lightly over medium heat, seasoning them with the *sambal oelek.*

- Mix the cooled rice with the fried mixture and divide evenly into eight portions, rolling each into a ball. Place each ball in the middle of a wonton wrapper; using a bit of egg white, seal the edges together while gathering the corners to make a pouch. Set aside until ready to deep-fry all the assembled dim sum.

- Prepare the "candy wraps": Heat the nut oil in a wok and fry the shallots, garlic, and shiitakes briefly over high heat. Season with the curry powder and soy sauce, remove from the heat, and allow to cool.

- Heat the sunflower oil in a small skillet and fry the grasshoppers for 1 to 2 minutes over medium-high heat.

- Divide the mushroom mixture among the wonton wrappers and place a grasshopper on each. Roll up the wrappers, sealing them with a bit of egg white, and twist the ends like a candy wrapper. Set aside until ready to deep-fry all the assembled dim sum.

Dim Sum. (Floris Scheplitz)

TRIANGLES

1 tablespoon sesame oil
½ celery rib, diced
1 spring onion, sliced
2 tablespoons sherry
2 teaspoons finely
    chopped fresh
    cilantro
¼ teaspoon ground
    cumin
1 tablespoon
    mealworms
Salt and freshly ground
    white pepper
8 wonton wrappers
Small amount of egg
    white

- Prepare the triangles: Heat the oil in the wok and stir-fry the celery, onion, sherry, cilantro, cumin, and mealworms over high heat until the mixture is tender and the liquid has completely evaporated. Season to taste with salt and white pepper and remove from the heat.
- When the mixture has cooled, divide evenly among the eight wonton wrappers. Using egg white to seal the edges, fold the wrappers over the filling to make little triangular shapes. Set aside until ready to deep-fry all the assembled dim sum.
- Heat oil in a deep fryer to 350°F (180°C). Fry the assembled dim sum in batches for 5 minutes, until nicely browned and crisp. Drain on paper towels and serve hot.

YIELD: 24 PIECES OF DIM SUM

Bitterbug Bites. (Floris Scheplitz)

Bitterbug Bites

Equipment needed: deep fryer, with vegetable oil for frying

RAGOUT FILLING

3 tablespoons (50 g) unsalted butter
¼ cup (60 g) all-purpose flour, plus a little extra for rolling
1¼ cups (300 ml) cold vegetable stock (homemade or prepared from a bouillon cube)
¾ ounce (20 g) mealworms
Salt and freshly ground black pepper

BREADING

3 egg whites, lightly beaten
7 ounces (200 g) bread crumbs

This recipe is based on a staple Dutch cocktail snack called *bitterballen*, which is round like a ball, but not at all bitter!

- Prepare the filling: In a lightly oiled nonstick pan, toast the mealworms for 1 or 2 minutes over medium heat until lightly browned. Remove from the pan and mince.
- Melt the butter in a small pan over medium-low heat and stir in the ¼ cup of flour with a spatula (this creates a roux).
- Cook the roux gently for 2 more minutes. Add half of the stock, whisking until smooth; then add the rest of the stock and cook the filling gently for 10 minutes, stirring frequently to prevent burning.
- Add the mealworms and season with salt and pepper. Pour the filling onto a platter, cover with plastic wrap, and allow to cool.
- Refrigerate the filling for at least 2 hours, until firm.
- Divide the filling into fifteen equal portions; roll these into balls, using a little flour to prevent sticking.
- Refrigerate the balls for 30 minutes. When firm, dip each in the egg white and then roll in the bread crumbs. Allow to firm up again in the refrigerator and cover again in the egg white and bread crumbs.
- Heat oil in a deep fryer to 350°F (180°C). Deep-fry the bitterbug bites in batches. Drain on paper towels.

YIELD: 15 BITES

Tips

- Substitute buffalo worms for the mealworms.
- Use this recipe to make five croquettes or rolls, instead of the fifteen bitterbug bites.

Fried Wontons. (Floris Scheplitz)

Bugsit Goreng (Fried Wontons)

Equipment needed: deep fryer, with vegetable oil for frying

1 tablespoon sesame oil
1 shallot, minced
1 garlic clove, crushed
4 tablespoons very finely sliced leek
½ cup (20 g) mealworms, coarsely chopped
1 teaspoon *sambal oelek* (raw hot chili paste), or more to taste
½ pound (230 g) ground meat (all beef or mixed pork and beef)
2 tablespoons bread crumbs
1 large egg
1 tablespoon soy sauce
Salt

20 (4-inch [10 cm] square) egg-roll wrappers
1 tablespoon all-purpose flour
Chili sauce

- Heat the sesame oil in a skillet over medium-high heat and fry the shallot, garlic, leek, mealworms, and *sambal oelek* lightly for about 2 minutes. Remove from the heat and allow to cool in the pan.
- Add the ground meat, bread crumbs, egg, and seasonings and knead well. This should be a well-seasoned mixture; add a little salt and more *sambal oelek*, if necessary.
- Whisk the flour into 3 tablespoons of water to make a paste.
- Divide the meat mixture evenly into twenty portions and roll each into a ball. Wrap each ball in an egg-roll wrapper and use the flour paste to seal the edges. Shape into a little pouch by gathering and twisting the corners together.
- Heat oil in deep fryer to 350°F (180°C). Fry the wontons in batches for 5 minutes, until nicely browned and crisp. Drain on paper towels and serve hot, with chili sauce for dipping.

YIELD: 20 WONTONS

Tips

- If you are using fresh mealworms instead of freeze-dried, use ¼ pound (120 g) of mealworms and ¼ pound (120 g) of ground meat.
- Rinse the mealworms and blanch them in boiling water for 1 minute. Rinse with cold water and drain well, then chop them coarsely, using a food processor. Knead with the remaining ingredients.

Mini Spring Rolls. (Floris Scheplitz)

Mini Spring Rolls

Equipment needed: deep fryer, with vegetable oil for frying

MARINADE

1 garlic clove, crushed
2 tablespoons peeled and grated fresh ginger
½ red chili pepper, seeded and finely chopped
Grated zest and juice of ½ lemon
2 tablespoons peanut oil
2 tablespoons Asian fish sauce
¼ cup (60 ml) coconut milk

FILLING

16 grasshoppers, legs and wings removed
½ ounce (15 g) dried rice noodles
2 tablespoons sunflower oil
1 onion, finely chopped
⅓ cup (50 g) very finely sliced green cabbage
3 tablespoons very finely sliced carrot
⅓ cup (50 g) bean sprouts

1 tablespoon all-purpose flour
8 (4-inch [10 cm] square) egg-roll wrappers
Chili sauce

Preheat the oven to 275°F (140°C).

- Prepare the marinade: With a mortar and pestle, grind the garlic, ginger, chili pepper, and lemon zest into a paste; add the lemon juice, oil, fish sauce, and coconut milk and stir until smooth.
- Prepare the filling: Place the grasshoppers in an ovenproof dish and cover with half of the marinade. Bake for 5 minutes.
- Bring 1 cup of water to a boil; turn off the heat and add the rice noodles. Allow to soak for 10 minutes. Pour the noodles into a colander, rinse with cold water, and drain well.
- Heat the sunflower oil in a wok over high heat and stir-fry the onion lightly. Add the cabbage and carrot, continuing to stir-fry briefly. When the vegetables have softened, add the bean sprouts and noodles.
- Season with the rest of the marinade; remove from the heat and allow to cool.
- Whisk the flour into 3 tablespoons of water to make a paste.
- Place one-eighth of the filling mixture and two grasshoppers on each egg-roll wrapper. Brush the edges with the flour paste and carefully (the wrappers tear easily!) wrap into a spring roll, tucking in and sealing the edges as you roll.
- Heat oil in deep fryer to 350°F (180°C). Deep-fry the spring rolls for 5 minutes, until nicely browned and crisp. Drain on paper towels and serve hot, with chili sauce for dipping.

YIELD: 8 SPRING ROLLS

Insects at a market in Laos. (Arnold van Huis)

Is It Healthy?

Contestant in an insect-cooking competition in Laos. (Thomas Calame)

MARGOT CALIS, INSECT FARMER

Fish Friday, Meat Loaf Wednesday, Insect Tuesday

Nestled in the meadows behind a white farmhouse lies Kreca insect farm, the business run by Margot Calis; her husband, Hans; and their daughters. Several new warehouses have been put up recently; indoor temperatures are tropical. Things are buzzing. Towering piles of insect-filled crates are lined up in rows: mealworm alley, wax moth alley, cricket alley . . .

The origin of Kreca dates back thirty years. The farm actually started out as a hobby that got a little out of hand. Margot Calis explains: "I have always been interested in breeding birds, and Hans raised fish. He once built a 'paludarium': an aquarium combined with plants, for amphibians. We kept frogs and lizards in it. Back then, Hans began his biology studies and I took care of the animals. I would use a net to collect food for them in the fields. But when it rained, I couldn't find any insects and had to get them elsewhere."

There had to be a better way. Calis asked her husband to help her set up a cricket colony, so that she could develop her own insect production. She started very modestly with a few crickets she obtained in Rotterdam, from a man who grew them in a small cupboard to feed his reptiles. Shortly thereafter, Calis met someone who normally had to import crickets from Switzerland, and he bought up her entire stock. This gave her the necessary incentive to get serious about her business in 1976.

"We moved in 1977. Hans built a 130-foot [40 m] square rearing room next to our new house. This space was already too small after two or three years. Because things were going so well, we decided to expand, and we bought this farm, where we have been ever since. In those days, Hans still worked as a zoology lab technician, and was also still studying. I spent my time rearing insects."

The new farm saw the start of the new company, Kreca: crickets (Dutch, *kre*kels) from *Ca*lis. Things got very busy, and Hans joined the business in 1980, as there was more than enough work for two. He started up a rearing of mealworms. The first few years, Kreca supplied insects mainly to a reptile importer, who sold the insects as feed, along with the reptiles. There turned out to be a good market for their products in the Netherlands, Belgium, and Germany. Calis says proudly:

"We were the first ones on the market within the Netherlands, and possibly even beyond its borders."

THE TURNAROUND: FARMING FOR HUMAN CONSUMPTION

The new business grew, with ups and downs. By 1986, they were producing crickets, yellow mealworms, and buffalo worms, all as animal feed. Farming insects for human consumption came later—actually very recently, around 2006 or 2007. "It started out with 'grasshopper man' and hunter Ruud Meertens. He considered grasshoppers a great product that could very well be eaten by people. Insect farmers Van de Ven and Kreca were immediately interested. Poulterer Jan Ruig was willing to join in the venture if he could offer several different products for sale; he thought one kind of insect was not enough. The three insect-rearing companies thus hatched the plan to start producing an assortment of three insect species, all suitable for human consumption."

Government authorities monitor the conditions under which the insects are reared. "Our farming procedures are designed to adhere to all NVWA [Netherlands Food and Consumer Product Safety Authority] regulations. We also buy only guaranteed and quality-controlled feed for the insects, from a certified company."

Calis feels that future prospects for producing insects for human consumption are all about two separate issues: her own business and market outlets. She doesn't see these two aspects as inextricably connected, because at present, her business produces mainly insects for animal consumption and only a very small amount for human consumption. "But I am sure that the latter will take off. We believe in slow and steady growth and development: that is how we work. Right now, producing insects is costly in terms of time and manual labor. If we hope to farm on a larger scale and contribute to the protein supply in the food chain, we will have to expand and mechanize."

INSECTS: REARING CYCLE

"We raise a total of fourteen species of insects, of which the buffalo worm is for human consumption. Van de Ven produces yellow mealworms, and the grasshopper specialist Ruud Meertens produces the migratory locust.

The buffalo worms go through the full life cycle. The small beetles lay their eggs in the food medium, and the larvae that hatch from the

Marieke Calis at work in the mealworm-rearing room. (Lotte Stekelenburg)

eggs grow there for about four weeks. We sift out the adult insects every week, and keep each weekly larval production separate. After four or five weeks, the larvae are sieved out of their food medium again, and they are ready for sale.

The insects are fed dry feed, such as grains, and fresh carrots. Mealworms are grown differently; they need to be moved onto fresh food regularly. Finally, the largest of the larvae are sifted out and are ready for consumption. There is a very nice little film on our Web site, showing how this works." The locusts are fed differently; they are raised on grass.

INSECTS: PROCESSING

"Insects used as animal feed are sold live. But we sell the buffalo worms for human consumption freeze-dried. We try to kill the insects in the most natural way possible, by cooling them slowly and gradually. We first refrigerate them, then put them in the freezer, so that they eventually reach a state resembling diapause, or winter sleep. In nature, too, insects die when it freezes."

Once frozen, the buffalo worms are dried by placing them in a machine that extracts all the moisture from them, in a vacuum. "We use the technique of freeze-drying because of food safety. The insects keep well like this; you can store them in a cupboard at home, and nothing can happen to them during transport."

INSECT TUESDAY

The Calis family does not eat insects every day. But sometimes, for special occasions, they will toss a bunch into a pan. "Next week we are celebrating the opening of our new building, so I will be serving up something. But this is more for promotional purposes, to acquaint people with insects and have them taste what they are really like. We don't eat insects every week, but occasionally, when I think of it, I throw a few buffalos into the wok; we eat them right up, just as we would eat nuts.

I remember, a long time ago, there used to be advertisements for 'Fish Friday' and 'Meat Loaf Wednesday.' Who knows; maybe one day we will be able to say, 'Insect Tuesday.'"

"A World That Works"

"My work focuses on social issues. A question I often ask myself is: How can you—sustainably—give everyone enough to eat and an adequate place to live? In other words, how do you create a world that works?" Marian Peters is an innovator; with Venik (Verenigde Nederlandse Insectenkwekers [Dutch Insect Farmers Association]), she is putting edible insects, as a new sector, on the map.

It must have been in 2006 or 2007, on a cold winter morning, when Marian Peters first heard about insects that you could eat. "A hunter, Ruud Meertens, was sitting in my garden, talking about grasshoppers and locusts. He knew that there was already a restaurant in Den Bosch that had insects on the menu." Peters and Meertens wondered about the potential for putting locusts on the market as a food-grade product for human consumption. Peters, who had worked on agricultural issues before, thought it would be economically feasible: "I thought: that could very well be a solution for all the empty pig barns here in Brabant" (a southern province in the Netherlands). Back then, the pork industry wasn't doing well, so she envisioned farmers making the switch from the curly-tailed to the six-legged. Meertens quite liked the idea of grasshoppers on wholesalers' shelves. "He asked me, 'Can't you find a subsidy for that?'" Peters accepted the challenge.

INCONVENIENT TRUTH

Desk research proved that Peters probably chose the right moment to tackle this subject. "I thought: there must be a fund with money for this; so I had a look around. It turned out that for years, the entomologists at Wageningen University had been proclaiming that we should be eating insects. The Food and Agriculture Organization of the United Nations [FAO] released a report in 2006 on the negative environmental effects of conventional meat consumption. To make a long story short: this couldn't go on any longer. Al Gore launched his film *An Inconvenient Truth*, and Marianne Thieme of the Dutch political party called the Party for the Animals released the film *Meat the Truth*. It all came together."

Peters talked to locust farmer Ruud Meertens and poulterer Jan Ruig. With them, she charted the entire production process needed to put food-grade insects on the market.

YOU'LL NEVER BE ABLE TO DO IT

Peters could tell right away that she had struck gold with this project. "Insects turned out to be *the* solution for the global hunger problem." Yet not everyone was enthusiastic about stocking insects in the stores. "But as soon as someone says to me, 'You'll never be able to do this,' I start going full speed. It will take time, though, before we are all eating insects. We have to be honest about that."

MEALWORMS

"Right now, one farm produces 3,300 pounds [1,500 kg] of beetle larvae a week, of which only a very small proportion is for human consumption. If we want to replace one-half of 1 percent of the total meat consumption with insect protein by 2020, we need to produce 8.15 million pounds [3.7 million kg] of insects. People need time to learn to eat insects—we can use that time to scale up."

INSECTS AS FOOD AND FEED

The biggest stumbling block is the insects' image problem. Our whole mind-set is that insects are considered disgusting. If you played with a daddy longlegs as a child, your mother would immediately exclaim, 'Yuck!'

"Insects can be put on the market along two pathways," the innovator continues, "as feed or as food." *Feed* is for animals. "Insects are an incredibly good substitute for fish oil and fish meal, which is becoming scarcer and thus more expensive. By using insects for feed, you help keep us from overfishing the oceans." *Food* means eating the insects ourselves. "You eat them whole, freeze-dried, or, possibly in the near future, fresh. Or you grind up the insects first, so you can make products with insect extract. My dream is an insect disassembly line; you put the insects in and take out the protein and the oil with its fatty acids. The skeleton, with its chitin, is suitable for industrial use."

VEGETARIANS

Peters would like to see vegetarians eat insects as well. She remembers the vote for the most delicious insect snack. "It was a battle between ground chicken meat with mealworms and a granola bar with mealworms. The granola bar won because so many vegetarians voted." So not every vegetarian is against it.

As an ingredient in meatballs, too, people seem to like buffalo worms and mealworms. "Recently, we ran a blind test with the enhanced meatball. The result was astounding: they made the meatball better. Suddenly, insects become an ingredient. The explanation is that insects have a nutty, whole wheat taste and texture, so they provide a good bite."

BUQADILLA

"Now we're working with various organizations to develop a so-called *buqadilla*, a falafel-like product with Mexican spices, containing 40 percent buffalo worms. It has been tested in three different company restaurants and has been well received. The *buqadilla* tastes good, but how do we sell it? As sustainable, healthy, or exotic?"

RECIPES

Peters ducked into the kitchen herself, just putting mealworms and locusts in a skillet. "It's fun to discover that it's best to fry locusts in very hot oil, such as hot pepper oil or nut oil. It gives them a delicious flavor and turns them into a great snack. It's better not to fry mealworms over such high heat, so butter is more suitable for those." To get people used to insects, Peters made mixtures with nuts, such as *katjang pedis*—a spicy peanut snack—with toasted mealworms; or she froze fresh mealworms, chopped them up with a food processor, and added them to meatballs.

Nutty Mealworms. (Lotte Stekelenburg)

Nutty Mealworms

½ cup (20 g) freeze-dried mealworms
8 ounces (225 g) spicy nut mixture, such as *katjang pedis*
(a spicy Indonesian peanut mix) or any spicy nut, seed, or
trail mix

- In a dry skillet, fry the mealworms over medium heat for
  no longer than a couple of minutes, stirring frequently and
  watching carefully, as they scorch easily.
- Allow to cool, then mix the meal worms with the spicy nut
  mixture. Enjoy!

Quick Meatballs. (Lotte Stekelenburg)

Quick Meatballs

¾ cup (150 g) fresh-frozen buffalo worms or mealworms
¾ pound (350 g) mixed ground pork and beef
1 (1.25 ounces [35 g]) packet meatball seasoning mix with
**   bread crumbs**
1 large egg
3 tablespoons (50 g) unsalted butter

- Grind the frozen insects in a food processor. In a bowl, knead the insects with the meat, seasoned bread crumbs, and egg until well combined. Roll the mixture into 1-inch (2½ cm) balls.
- Allow the butter to brown in a Dutch oven over medium-high heat and brown the meatballs, turning them frequently. Simmer, covered, for up to 20 minutes, until cooked through. If needed, add a little water to the pan, to prevent scorching.

YIELD: SERVES 4

Oatmeal Bars. (Lotte Stekelenburg)

Oatmeal Bars

Equipment needed: cookie sheet lined with parchment paper

⅔ cup (150 g) unsalted butter
1¼ cups (200 g) rolled oats
⅔ cup (25 g) freeze-dried buffalo worms
¼ cup (60 g) raisins or dried cranberries
¾ cup (150 g) sugar
1 teaspoon baking powder
1 large egg
2 tablespoons all-purpose flour
Pinch of salt
Zest and juice of 1 lemon

Preheat the oven to 350°F (180°C).

- Melt the butter in a small pan over low heat.
- Combine all the other ingredients in a bowl. Add the melted butter and mix well. Chill for 1 hour in the refrigerator. Roll out the dough onto the prepared cookie sheet, and shape into a rectangle about ½ inch (12 mm) thick.
- Bake for about 20 minutes, until lightly browned. Cut into bars while still warm, then allow to cool.

YIELD: 16 BARS

Eating Insects Safely

There are about 6 million insect species on Earth, of which at least 1,900 are eaten. Not all insects can be safely eaten, however. Many of them feed on plants, and quite a few plants produce toxic substances to defend themselves against enemies such as herbivorous insects. Insects that are resistant to the plant toxins can eat the plants without ill effects, and as they do, they may accumulate these toxins in their body. This often results in higher concentrations in the insects than in the plants themselves. Special care should be taken to render such insects safe for consumption.

In South Africa, there are some very poisonous true bugs, related to aphids and cicadas. The bugs excrete a brownish liquid that can stain our hands and be harmful to our eyes. Before eating these bugs, they must be washed a few times in warm water to rinse out the poisons; this turns the bugs from green to yellow. The water used for washing the bugs is so toxic that it can be used as a pesticide to kill other insects. The rinsed bugs are then cooked with a little salt and eaten as a snack. Another example of a toxic insect is a very colorful grasshopper in West Africa, known as the "stinking" grasshopper. Once the glands that produce the smell have been removed, the grasshopper can be safely eaten. There are also poisonous insects in the rest of the world, so it is not a good idea to catch just any insect in your garden and make a meal of it. You should be especially wary of brightly colored insects, as they are often toxic. Because birds know that, they often steer clear of colorful insects.

Stinking grasshopper (*Zonocerus variegatus*). (Christiaan Kooyman)

Another risk inherent to eating insects is poor hygiene during their handling. They can become infected with harmful bacteria, for example. This risk of contamination is also present in the tropics, where caterpillars are often just placed outside on a piece of plastic sheeting, to dry in the sun.

Insects meant for human consumption, then, must come from a reliable source. In the Netherlands, for example, insect producers are required to conform to all existing regulations for food production, just like any meat producer. The buffalo worms, mealworms, and locusts that are sold by the Dutch and American suppliers (see page 179) are safe. They are grown carefully under hygienic conditions and are regularly tested for quality, just like any other foods.

What Kinds of Insects Can Be Eaten?

Most insects that are eaten belong to one of four major groups:

- Beetles
- Hymenoptera, such as ants, bees, and wasps
- Caterpillars
- Grasshoppers, locusts, and crickets

Beetles make up the largest percentage of edible insects: about 31 percent of the more than 1,900 species eaten are beetles. This, of course, is not surprising, as about 40 percent of all insect species in the world are beetles. This group includes mealworms. In Europe, these beetle larvae are mostly grown for animal feed, but they are easy to grow and very suitable as a human food as well.

Ants and wasps are another important group. In Laos, people eat the pupae of weaver ants, and in other parts of Southeast Asia, wasp and honeybee larvae are popular. Caterpillars are especially favored

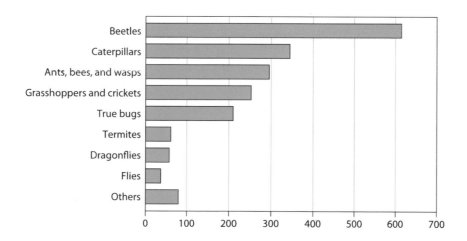

Number of edible insect species per group of insects (total = 1,959).
(Yde Jongema, "List of Edible Insects of the World," April 2012, http://www.
wageningenur.nl/en/Expertise-Services/Chair-groups/Plant-Sciences/Laboratory-
of-Entomology/Edible-insects/Worldwide-species-list.htm)

Market in Chiang Mai, Thailand. (Arnold van Huis)

in central and southern Africa, whereas crickets and grasshoppers are widely eaten both in Africa and in Asia. The majority of insects eaten are collected in the wild, but in Southeast Asia, crickets are also reared for people to eat.

Spiders have eight legs and are therefore technically not insects. They do belong, however, to the larger group of arthropods, which includes insects, lobsters, and shrimps. Because they are so closely related to insects, and the sensation of eating spiders is comparable to that of eating insects, we have included them in this book. Several species of spiders are on the menu in Africa, Asia, and South America.

Insect Consumption and Health

It is not easy to make generalizations on the nutritional value of insects, simply because there are so many different edible species. The taste and nutritional value of insects depend on their developmental stage (for example, larva or adult) and on what the insects themselves have eaten. As with most foods, the preparation and processing methods (drying, boiling, or frying, for example) preceding consumption will also influence the insects' nutritional composition. The main nutritional components of insects are proteins, fats, and fibers.

The nutritional value of proteins depends on their content, quality (essential or nonessential amino acids), and digestibility. Insect protein content ranges between 20 and 75 percent; for beef, between 40 and 75 percent. Insects are a good source of essential amino acids. Whereas cereal proteins are often low in lysine (1 to 4 percent) and, in some cases, lack the amino acids tryptophan (for example, corn) and threonine, these amino acids are very well represented in several insect species. For example, several caterpillars, palm weevil larvae, and aquatic insects have lysine contents of more than 5 to 9 percent of crude protein. The digestibility of insect proteins is generally high, and comparable to that of beef: 78 to 99 percent of insect protein is absorbed by the human body.

Edible insects are generally a considerable source of fat. Caterpillars, palm beetles, and termites have a fairly high fat content. Their oils are rich in polyunsaturated—the good—fatty acids and frequently contain the essential fatty acids linoleic and alpha-linolenic acid. The nutritional importance of these two fatty acids, especially for the healthy development of infants and children, is well recognized.

Edible insects are also rich in minerals. They often contain more than the 6 milligrams per 100 grams of iron in beef. For example, the range of iron content per 100 grams of dry weight of the mopane caterpillar is between 31 and 77 mg; and that of the migratory locust, between 8 and 20 mg. Some crickets even contain more than 1,500 mg. This high iron content is invaluable for people in developing countries, where about half of all pregnant women and up to 40 percent of preschool children suffer from anemia. In general, most insects are also good sources of zinc. Zinc deficiency is the fifth-highest risk factor for disease in developing countries and can lead to diarrhea, skin lesions, chronic liver disease, chronic renal disease, sickle-cell disease, diabetes, malignancies, and other chronic illnesses. Whereas beef averages 12.5 mg of zinc per 100 g (dry weight), the palm weevil larva contains an impressive 26.5 mg; and the cricket, 25 mg per 100 g dry weight.

In general, edible insects provide more vitamins B_1 and B_2 than does whole wheat bread. There is also a fair amount of vitamin B_{12} in mealworm larvae, mopane caterpillars, and house crickets, but not in many other edible insect species. Although some caterpillar species contain vitamin A, most insect species are not a good source of this nutrient. Vitamin E features in palm weevil larvae and in silk worms.

The most common form of fiber in insects is chitin. This insoluble component is found in the insect skeleton, as it is in crabs, lobsters, and shrimps. The chitin content of insects varies between 0.25 and 5 percent of their total fresh weight. Some people have an enzyme in their body that can digest chitin. This enzyme is more prevalent in tropical countries—where insects are regularly consumed—than in Western countries. Chitin has been associated with defense against parasitic infections and some allergic conditions.

In short, how does the composition of mealworms compare with that of beef?

- *Overall*. The fat content of beef is higher than that of mealworms. Beef is slightly lower in moisture content than are mealworms and is marginally higher in protein and metabolizable energy.
- *Minerals*. Mealworms and beef contain comparable values of copper, sodium, potassium, iron, zinc, and selenium.
- *Amino acids*. Beef is richer in glutamic acid, lysine, and methionine and has less isoleucine, leucine, valine, tyrosine, and alanine than do mealworms.
- *Vitamins*. Mealworms have generally higher vitamin concentrations than does beef, with the exception of vitamin B_{12}.
- *Fatty acids*. Beef contains more palmitoleic, palmitic, and stearic acid than do mealworms, but the essential linoleic acid is far higher in mealworms.

This comparison allows us to conclude that the nutritional value of mealworms is comparable to that of beef.

Insect buffet served during the FAO conference "Edible Forest Insects: Humans Bite Back!" held in Chiang Mai, Thailand, in 2008. (Arnold van Huis)

FIVE APPETIZERS

Flower Power Salad. (Floris Scheplitz)

Flower Power Salad

DRESSING

2 tablespoons olive oil
2 tablespoons nut oil
3 tablespoons red wine
    vinegar
1 teaspoon Dijon
    mustard
1 teaspoon honey
Freshly ground black
    pepper
Sea salt

SALAD

4 ounces (125 g) mixed
    lettuces, such as leaf
    lettuce, corn salad,
    arugula, radicchio,
    and frisée
1 handful of colorful,
    edible flowers, such
    as roses, violets,
    garden nasturtiums,
    borage flowers,
    daylilies, and dahlias
1 tablespoon olive oil
12 grasshoppers, legs
    and wings removed
⅔ cup (25 g)
    mealworms
Salt and freshly ground
    black pepper

- Mix the dressing ingredients well in a dressing shaker.
- Prepare the salad: Wash and dry the lettuce, and tear any large leaves into bite-size pieces. Arrange them artistically on a large platter, then drizzle with some of the dressing.
- Drape the flowers over the lettuce. Of the large flowers, such as roses and dahlias, use only the petals.
- Heat the oil in a nonstick pan over medium-high heat and fry the insects for about 2 minutes, until lightly browned. Season with salt and pepper and scatter them over the salad.
- Serve the rest of the dressing separately.

YIELD: SERVES 4

Tips

- Instead of using a dressing shaker, you can use an empty, clean jam jar with a lid.
- Edible flowers are available in specialty shops, or picked from your own garden.
- Make sure the flowers are organic and clean.

Thai Salad. (Floris Scheplitz)

Thai Salad

Juice of 1 lime
1 tablespoon Asian fish
    sauce
1 tablespoon soy sauce
1 tablespoon sesame oil
1 garlic clove, crushed
1 hot red pepper,
    seeded and minced
1 teaspoon ground
    cumin
1 tablespoon stem
    ginger syrup

DRESSING

2 tablespoons rice
    vinegar
3 tablespoons soy sauce
1 tablespoon tomato
    ketchup
1 tablespoon white wine
4 tablespoons
    sunflower oil
1 shallot, minced
1 (¾-inch [2 cm]) piece
    fresh ginger, peeled
    and grated

SALAD

20 grasshoppers, legs
    and wings removed
2 tablespoons peanut,
    safflower, or canola oil
2½ ounces (75g) mixed
    lettuces with arugula
2 tablespoons sesame
    seeds, toasted
2 tablespoons chopped
    fresh cilantro leaves

- In a blender, puree the marinade ingredients until smooth.
- Mix the dressing ingredients in a dressing shaker (or clean jam jar with a lid), shaking thoroughly.
- Prepare the salad: Place the grasshoppers in a small bowl and pour the marinade over them. Marinate for 30 minutes and then allow them to drain in a sieve placed over a bowl.
- Heat the wok oil in a skillet over medium-high heat, and fry the grasshoppers until browned.
- Toss the lettuce in a bowl with the dressing and divide among four plates.
- Arrange the fried grasshoppers on the salad and sprinkle with the sesame seeds and cilantro leaves. A little bit of marinade poured over the salad is a delicious finishing touch.

YIELD: SERVES 4

Vegetable Carpaccio

Equipment needed:
vegetable slicer or
mandoline

CARPACCIO

4 waxy potatoes (may
  use red-skinned)
4 broccolini (baby
  broccoli) florets
1 medium-size carrot,
  peeled
8 radishes
¼ head frisée
Fresh herbs, such as
  chervil, chives, and
  parsley
⅔ cup (25 g) buffalo
  worms

ONION CREAM

2 onions, finely
  chopped
2 tablespoons safflower
  oil
2 tablespoons white
  wine

DRESSING

3 tablespoons white
  wine vinegar
1 tablespoon stem
  ginger syrup
6 tablespoons safflower
  oil
¼ teaspoon Dijon
  mustard
Salt and freshly ground
  white pepper

- Prepare the carpaccio: Boil the unpeeled potatoes in plenty of water for 15 minutes, until just done. Leave in the cold water to cool.
- Slice the broccolini florets twice lengthwise and blanch them for 30 seconds in boiling salted water. Cool them right away under cold running water.
- Using the vegetable slicer or mandoline, thinly slice the carrot, radishes, and cooked potatoes.
- Wash and dry the frisée and herbs, discarding any yellowed leaves. Gather up attractive little bunches of chervil and parsley, and chop the chives finely.
- Toast the buffalo worms in a lightly oiled nonstick pan over medium-high heat for 1 or 2 minutes, stirring frequently until lightly browned.
- Prepare the onion cream: Heat the safflower oil in a small skillet and fry the onions over low heat. When the onions are soft and translucent—they should not brown—add the white wine, stirring constantly until it is absorbed, 1 to 2 minutes. Remove from the heat and puree the onions with a stick blender to make a smooth cream.
- Mix the dressing ingredients.
- Spread some of the onion cream onto each of four flat plates, using the back of a spoon.
- Divide the vegetables among the plates, making a colorful palette. Sprinkle the dressing over the vegetables and decorate the plates with herbs, flowers, and the roasted insects.

YIELD: SERVES 4

Vegetable Carpaccio. (Floris Scheplitz)

Edible flowers, such as borage or garden nasturtiums

Tips

- You can use thin asparagus spears, preferably raw, instead of the baby broccoli.
- If you prefer, you can blanch the carrot slices slightly to make them softer.

Pumpkin Soup. (Floris Scheplitz)

Pumpkin Soup

Equipment needed:
large roasting pan

2 pounds (1 kg)
    pumpkin
1 onion, coarsely
    chopped
1 garlic clove, thinly
    sliced
1 red chili pepper,
    seeded and sliced
    into strips
5 fresh sage leaves,
    cut into strips, or 2
    tablespoons dried
2 teaspoons *ras el*
    hanout (a Moroccan
    spice mixture, or use
    garam masala)
⅔ cup (160 ml) olive oil
Sea salt
Freshly ground black
    pepper
6 cups (1.5 L) vegetable
    stock (homemade
    or prepared from a
    bouillon cube)
1¼ cup (50 g)
    mealworms
½ cup (120 ml) crème
    fraîche or sour cream
6 tablespoons plain
    yogurt

Preheat the oven to 350°F (180°C).

- Wash and dry the pumpkin and remove the seeds and fibers, but do not peel. Cut the pumpkin into large chunks and place them in the roasting pan.
- Sprinkle the onion, garlic, chili pepper, sage, spices, olive oil, sea salt, and black pepper over the pumpkin chunks and mix well. Bake for 40 minutes.
- Transfer the contents of the roasting pan to a large saucepan and add the stock. Bring the soup to a boil, then puree with a stick blender.
- Toast the mealworms in a lightly oiled nonstick pan over medium-high heat for 1 to 2 minutes until lightly browned.
- Adjust the soup seasonings with sea salt and black pepper; thin with a little water, if desired.
- In a small bowl, combine the crème fraîche and yogurt.
- Divide the mealworms among eight bowls and ladle the hot soup over them. Finish with a dollop of the cream mixture.

YIELD: SERVES 8

Couscous Salad. (Floris Scheplitz)

Couscous Salad

8 dried apricots
1½ cups (250 g) dried
  couscous
Salt
4 tablespoons olive oil
Juice of ½ lemon
½ cucumber, diced
1 yellow bell pepper,
  seeded and diced
12 cherry tomatoes,
  halved
2 tablespoons fresh
  parsley, chopped
2 tablespoons, fresh
  mint, chopped
2 tablespoons fresh
  cilantro, chopped
2 spring onions, sliced
  into rings
2 tablespoons sliced,
  pitted black olives
1 tablespoon pine nuts
½ cup (20 g) freeze-
  dried buffalo worms
Freshly ground white
  pepper
½ lemon, quartered

MARINADE

1 tablespoon soy sauce
1 tablespoon stem
  ginger syrup
1 tablespoon olive oil
Pinch of cayenne pepper
1 teaspoon *ras el hanout*
  (a Moroccan spice
  mixture, or use garam
  masala)

12 grasshoppers, legs
  and wings removed
12 tablespoons olive oil

- Prepare the salad: Cover the dried apricots in warm water for 20 minutes to plump.
- Bring 1¼ cups of water and 1 teaspoon of the salt to a boil. Remove from the heat, sprinkle in the couscous, cover, and let stand for 5 minutes. Fluff the couscous with a fork and mix in the olive oil and lemon juice. Refrigerate the couscous for 30 minutes.
- Mix the cucumbers, yellow pepper, tomatoes, spring onions, olives, and herbs in a serving bowl. Drain the apricots, cut them into strips, and add to the bowl.
- Mix the marinade ingredients in a separate bowl and add the grasshoppers. Let marinate for 10 minutes.
- Drain the grasshoppers in a sieve. Heat the olive oil in a pan over medium-high heat and fry the grasshoppers for 2 minutes until crisp.
- Separately, toast the pine nuts and buffalo worms for 1 minute each in a lightly oiled nonstick pan over medium-high heat until lightly browned.
- Add the couscous, pine nuts, and buffalo worms to the vegetable mixture and mix gently. Add salt and freshly ground white pepper to taste.
- Garnish the salad with the grasshoppers and lemon sections.

YIELD: SERVES 4

Marcel Dicke eating a salad with mealworms and locusts. (Lotte Stekelenburg)

3

Eating Insects

NATURALLY!

Chapulines al ajillo. (Lotte Stekelenburg)

JAN RUIG, POULTRY AND GAME SPECIALIST

"Some People Won't Try Anything New"

There they are, humbly sitting on a little shelf to the right of the counter in Jan Ruig's store: small plastic containers of bugs. Almost too distracting is the impressive parade of pâtés, sausages, and hams, or such exotics as crocodile and kangaroo meat, all of which this poulterer also has in his assortment. And yet Ruig sees it as his mission to be a purveyor of locusts, mealworms, and buffalo worms.

"Don't forget that we are doing development work, though you could even call it missionary work." Jan Ruig doesn't mean to say that he is a do-gooder. But then again… "Jan Ruig can't make the world a better place, but I can help it along by bringing items that are eaten in other countries to the Western world, to the Netherlands. That benefits our gastronomic culture. I'm working on putting something new on the market, and on imparting my vision and philosophy regarding food. I don't think that we have to be eating insects en masse, but I do think that it can thrive here," the poultry specialist elaborates. Right now, not many in the Netherlands are yet sold on a crispy little mealworm or locust.

MEMORY

Neither was Ruig at first. It must have been around 2006 or 2007 when he was approached by a hunter he knew, Ruud Meertens. "He asked whether it wouldn't be a good idea to put locusts on the market." The poulterer, in a stage whisper: "I said, 'Insects? They're good only to swat.'" He laughs at the memory. "But I didn't say no immediately. It's a source of protein, and as a farmed product, it fits in with game and poultry, so I did want to give it some thought. A few days later, I called him to say that I would be willing to try it."

What made him change his mind? In the days between the seemingly strange suggestion from Meertens and the phone call, Ruig had consulted with his team. First with the management, and then with the guys in sales. "At first, most of them reacted just as I had. One laughed; another was enthusiastic right away. But at some point we just fried up some locusts. We made a little *bitterball* [a typical deep-fried Dutch snack], and the idea started to catch on. In the beginning it was a little creepy, but as soon as the members of the team had eaten them, the awkwardness disappeared. And we ate some again the next time."

Before long, Ruig and his team agreed that their resistance to insect consumption was all in their head. "Actually, pretty soon everyone was thinking, 'Why don't we have a closer look at this new source of protein?'"

FAMILY BUSINESS

Ruig is always bent on innovation. "What does a poultry and game dealer do? Sell meat taken from the wild. And where does an insect live? Right, in the wild. So that connection is obvious. We are always looking at what is or isn't allowed in the world of game. We also sell rattlesnake. And if something comes from the Netherlands, I am a complete believer in checking to see whether it can be sold. Our vision is: whatever game and poultry is available in the Netherlands should be put on the market. If there were no insects farmed in the Netherlands, I would have said no right away. It's also because we, as a family business, have to have an affinity with it. I didn't even know, by the way, that insects were being farmed in this country."

Because he was so curious, Ruig got himself invited to a locust farm. "I got a real eyeful there. I saw how sensitively they deal with the animals. And how they have tremendous knowledge about insects. And that on our own soil. Fantastic." He likes to compare them to chicken and cattle farmers. "These aren't people who are just working to put food on their plates; they are concerned with their animals and the animals' welfare." Later, he also went to see the mealworms and the buffalo worms. He was getting more enthusiastic all the time about the plan to put the freeze-dried beetle larvae and locusts on the market.

So there was no lack of supply. But what about the demand? To sell a product successfully, you need customers. "The issue back then was: How can I put this on the market safely? What do I need to do?" Ruig had his first success with Sligro; the wholesaler showed interest in selling insects. The poulterer knew that it would go somewhere, "in ten years." He was in no hurry. "You invest, money goes into it, time goes into it, and energy, and you know that it can be years before you can reap the fruits."

IT TOOK SOME DOING

In 2008, Ruig was able to convince the Koksgilde (a professional chefs' association) to cook with the little critters at Horecava, the annual food industry exhibition in the Netherlands. "That really took some doing. Even the RAI, the Amsterdam convention venue, had to give us permission. It was so different from what people were used to." Insects were incorporated into bonbons, *bitterballen*, quiches, and more.

"Ten thousand morsels. And they were polished off. Unprecedented!" An anecdote: "We had bonbons made, with a locust on top and a buffalo worm inside. Some guests gingerly removed the locust and proceeded to enjoy the bonbon. Not knowing that there was also an insect inside. We asked them, 'Do you know what you have eaten?' 'Gee, delicious, what an unusual taste,' was the reaction. And finally: 'Why not?' Young people especially like to try it. And people who have traveled a lot to other countries. They know that insects are eaten there."

Next, Ruig did—if you can call it that—market research, using trial and error. "Just by trying it out. What can you do with it? And our sales representatives went on the road with containers of bugs to survey our customers: 'Is this something for you?'" In addition, Ruig headed into the kitchen, to cook and to taste. With insect chef Henk van Gurp, he developed such dishes as insect *bitterballen* (Bitterbug Bites [see page 47]). And he discovered that he likes to stir-fry locusts, and mix buffalo worms into his salads like croutons or fried bacon bits. Nice and simple. You could eat them raw, but the advice is to go ahead and fry them up. Or even smoke them. "Really different, not like smoked eel, but incredibly delicious. Hard to describe. Nutty. Always nutty. If you don't like nuts, you won't like insects."

ATTENTION

"Some people won't try anything new, and that's not going to change. But I would really like to show what the possibilities are for eating insects. You can't force acceptance; it has to grow." A handful of adventurous restaurants have put the critters on the menu. "Especially where the chefs have direct contact with the guests, selling insects works well." Consumers usually order via the Ruig Internet shop (see page 180), though people in the neighborhood just come in to pick up ten jars of locusts. And now everyone is talking about eating insects. Great, but Ruig isn't in it for the attention. And what if everything is a flop, and nobody is going to eat insects? "Then it's just our bad luck." But still, to the poulterer, selling insects is more than just an experiment. "I think that, in the future—not tomorrow, but in a few years—the world will have to contend with a shortage of protein sources. Part of that can be solved by eating insects."

LAND SHRIMP

Perhaps we will shift to very different eating habits in the coming years. Ruig follows culinary developments very closely. "I have been in this business for more than forty-five years, and if I look at how we used to eat and now, there's almost no comparison. Back then, whole chickens went to the restaurants, and the chef made something with them; these days, we sell them fried, smoked, as nuggets, every which way. We used to eat only three times a day, and now ten times. Whenever we want. And we get food 'out of the wall,' from an automat. I still remember when the first Febo automat opened, when I was eighteen. Everyone thought that was crazy. Take a look around now. Who *doesn't* eat that way?" Yet Ruig doesn't foresee a fast-food grasshopper. "A locust makes a beautiful decoration, and is ideal for snacking. Just fry it for a moment in oil with some spices. A delicacy." But not yet for mass consumption. "It is and will always be a niche market. It has a little something special. Some people call the grasshopper a land shrimp. That sounds very different, huh? But think about it: How long ago did we start eating shrimp in the Netherlands, with all those tentacles and things?" The poulterer has every confidence that arthropods from the land or out of the sky will become just as ordinary as North Sea shrimp.

And for him? Personally, Ruig likes to munch on a portion of "land shrimp." As a first course. Followed by a main course with, oh yes, a good piece of meat.

Spicy "Land Shrimp" Snack. (Lotte Stekelenburg)

Spicy "Land Shrimp" Snack

2 tablespoons hot pepper oil
25 locusts or grasshoppers, legs and wings removed
Juice of ¼ lime
Salt

- Heat the oil in a skillet over medium-high heat and fry the insects for a few minutes, until lightly browned.
- Sprinkle with the lime juice and continue to heat over low heat until the juice has evaporated and the "land shrimp" are nice and crisp.
- Sprinkle with a little salt.

Tip

- You can use all kinds of herb salt mixtures to flavor these. Or, instead of lime juice, season with something like lemon pepper.

ELEVEN MAIN DISHES

Minestrone. (Floris Scheplitz)

Minestrone

3 tablespoons olive oil
2 garlic cloves, crushed
2 onions, finely chopped
4 ribs celery, sliced
2 carrots, diced
3 quarts (3 L) vegetable
  stock (homemade
  or prepared from a
  bouillon cube)
1 (15-ounce [400 g]) can
  diced tomatoes
Scant ½ cup (120 ml)
  red wine
1 (15-ounce [400 g]) can
  white beans (navy or
  cannellini), drained
  and rinsed
8 ounces (225 g) green
  beans, cut into bite-
  size pieces
2 zucchini, cut into
  chunks
3 ounces (85 g) spinach
  or Swiss chard,
  coarsely chopped
1 tablespoon chopped
  fresh basil
1 tablespoon chopped
  fresh oregano
3 ounces (85 g) dried
  pasta (broken
  spaghetti, for
  example)
⅔ cup (25 g) buffalo
  worms
¼ cup (10 g)
  mealworms
Sea salt and freshly
  ground black pepper
Small piece of Parmesan
  cheese, grated
Extra-virgin olive oil, for
  serving

- Heat the olive oil in a large soup pot over low heat. Add the garlic and onions and fry for a few minutes. Then add the celery and carrots and continue to fry for another 2 minutes.
- Add the stock, tomatoes, and wine and bring to a boil, stirring constantly. Add the white beans, green beans, zucchini, spinach, and herbs. Cover and simmer for 30 minutes over low heat.
- Add the pasta to the soup during the last 10 minutes of cooking time and simmer until al dente.
- Toast the insects for 2 minutes in a lightly oiled non-stick pan over medium-high heat and add them to the soup just before serving.
- Taste the soup and add sea salt and pepper, if necessary.
- Serve the Parmesan cheese and olive oil at the table, so that each person can add however much he or she likes.

YIELD: SERVES 4

Tagliatelle with Creamy Herb Sauce. (Floris Scheplitz)

Tagliatelle with Creamy Herb Sauce

SAUCE

6 tablespoons olive oil
1 onion, finely chopped
1 garlic clove, crushed
½ cup (120 ml)
    vegetable stock
    (homemade or
    prepared from a
    bouillon cube)
Scant 1 cup (220 ml)
    heavy cream
1 sprig fresh rosemary
2 sprigs fresh thyme
Salt and freshly ground
    black pepper
⅔ cup (25 g) freeze-
    dried mealworms
2 tablespoon fresh
    basil, chopped
1 tablespoon pine nuts,
    toasted

PASTA

1 pound (500 g) fresh
    tagliatelle
Salt
Freshly ground black
    pepper

- Prepare the sauce: Heat 3 tablespoons of the olive oil in a skillet over medium heat and sauté the onion and garlic until softened but not starting to brown. Add the stock, cream, and herbs and simmer gently until reduced by half, about 10 minutes.
- Prepare the pasta: While the sauce is simmering, cook the tagliatelle in boiling salted water until al dente. Drain the pasta and season with the rest of the olive oil and freshly ground pepper to taste.
- Toast the mealworms in a lightly oiled nonstick pan over medium-high heat for 2 minutes. With a fork, remove the herb sprigs from the sauce. Add the mealworms to the sauce and season to taste with salt and pepper.
- Divide the pasta among four plates and ladle the sauce on top. Garnish with the basil and pine nuts.

YIELD: SERVES 4

Ravioli

Equipment needed:
pasta machine

DOUGH

2½ cups (600 g) durum
  wheat flour
1¼ cups (200 g)
  semolina
1 teaspoon salt
5 large eggs
3 tablespoons olive oil

FILLING

2 tablespoons olive oil
1 onion, finely chopped
⅔ cup (25 g)
  mealworms
½ pound (250 g)
  spinach, washed and
  dried well
½ cup (100 g) ricotta
  cheese
Salt and freshly ground
  white pepper

SAUCE

4 tablespoons olive oil
1 large onion, finely
  chopped
2 garlic cloves, crushed
4 tablespoons tomato
  paste
1 (15-ounce [400 g]) can
  diced tomatoes

- Prepare the dough: Mound the flour, semolina, and salt directly on a clean kitchen counter. Make a well in the middle and pour in the eggs and olive oil. Mix well with your fingertips to incorporate the ingredients, then knead the dough for 10 minutes. If the dough is too dry, add a bit of water; it should become smooth and elastic. Divide the dough into four portions and shape each into a ball. Cover and allow to rest for 2 hours.
- Prepare the filling: Heat the olive oil in a wok over medium-high heat and stir-fry the onion and mealworms for about 2 minutes. Add the spinach and stir-fry until the spinach has wilted. Transfer this mixture to a sieve and allow to drain and cool. Transfer the spinach mixture to a cutting board and chop it roughly. Place in a bowl and mix in the ricotta and salt and pepper to taste.
- Prepare the sauce: Heat the olive oil in a saucepan over medium heat and fry the onion and garlic gently for 2 minutes. Add the tomato paste and cook for 2 more minutes. Then add the diced tomatoes, sugar, and oregano and allow to simmer for 30 minutes over low heat. Season the sauce with salt and pepper.
- Pat down the balls of dough on a floured board and run them through the widest setting of the pasta machine. Fold each sheet in half and run it through the machine again. Dust the sheets with flour so that they don't stick. Repeat two more times. Reduce the setting on the machine one step at a time and run the sheets of dough through about three more times, until they are about 1 mm thick.

Ravioli. (Floris Scheplitz)

1 tablespoon sugar
1 tablespoon fresh oregano, chopped, or 1 teaspoon dried
Salt and freshly ground black pepper

All-purpose flour, for dusting
2 tablespoons olive oil and 1 tablespoon salt, for cooking the ravioli
1 cup (100 g) grated Parmesan cheese

- Lay two sheets of dough on the counter and moisten them with a little water. Place little balls of filling about 2½ inches (6 cm) apart on the dough. Cover with the other two sheets of dough, pressing down well to remove all air bubbles. Cut the ravioli with a knife or a pastry wheel into 2½-inch (6 cm) squares.
- Cook the ravioli: Bring a large pan of water with 2 tablespoons of olive oil and 1 tablespoon of salt to a boil. When it boils, add the ravioli and cook them for about 8 minutes, until al dente. Drain and divide among four plates. Ladle the sauce over the ravioli and sprinkle the Parmesan on top just before serving.

YIELD: SERVES 4

Wild Mushroom Risotto. (Floris Scheplitz)

Wild Mushroom Risotto

RISOTTO

2 tablespoons olive oil
1 onion, minced
1 garlic clove, crushed
1¾ cups (400 g) raw arborio rice
¼ cup (60 ml) dry white wine
4 cups (1 L) vegetable stock (homemade or prepared from a bouillon cube)

MUSHROOM MIXTURE

3 tablespoons olive oil
2 tablespoons (10 g) unsalted butter
½ pound (220 g) mixed wild mushrooms, such as chanterelles, oyster, shiitake, porcini, and portobello, cleaned and chopped roughly
1 garlic clove, crushed
1 teaspoon chopped fresh thyme
⅔ cup (25 g) buffalo worms
12 grasshoppers, legs and wings removed
Salt and freshly ground black pepper

1 tablespoon red wine vinegar
4 tablespoons (50 g) unsalted butter
½ cup (50 g) grated Parmesan cheese
3 tablespoons chopped fresh parsley

- Prepare the risotto: Heat the olive oil in a saucepan over medium heat and fry the onion and garlic gently until soft but not starting to brown. Add the rice and fry gently for 1 more minute. Pour in the white wine and half of the stock and stir constantly while the rice continues to cook gently. When the liquid has been absorbed, add another splash of the stock, again stirring and cooking until absorbed. Repeat until the rice is tender and creamy, about 15 minutes.
- Prepare the mushrooms: Heat the olive oil and butter in a heavy-bottomed pan over medium-high heat. When hot, add the mushrooms and fry until nicely browned. Add the garlic, thyme, buffalo worms, and grasshoppers and fry for 1 or 2 minutes. Season to taste with salt and pepper and remove from the heat.
- Mix half of the mushroom mixture, the red wine vinegar, butter, half of the Parmesan, and half of the parsley into the risotto. Divide the risotto among four plates. Top with the remaining mushrooms. Garnish with the rest of the Parmesan and parsley.

YIELD: SERVES 4

Hakuna Matata. (Floris Scheplitz)

Hakuna Matata

2 tablespoons sake
   (Japanese rice wine)
½ cup (120 ml) soy sauce
1 tablespoon stem
   ginger syrup
1 tablespoon brown
   sugar

20 grasshoppers, legs
   and wings removed
½ cup (20 g) mealworms
½ cup (20 g) buffalo
   worms

1¼ cups (200 g) raw
   basmati rice
4 tablespoons sunflower
   oil
1 garlic clove, crushed
½ red chili pepper,
   seeded and finely
   chopped
1 (½-inch [1 cm]) piece
   fresh ginger, peeled
   and grated
4 ounces (100 g)
   mushrooms, sliced
2 ounces (50 g) snow
   peas
2 tablespoons (20 g)
   canned corn
2 spring onions, sliced
   diagonally into ½-inch
   (1.3 cm) pieces

This recipe was inspired by a song from *The Lion King.*
Hakuna matata is a Swahili term that means something
like, "Don't worry, be happy." The animals that sing this
song love insects: very fitting!

- Heat the marinade ingredients in a saucepan for 2
  minutes over low heat. Allow to cool. Add the insects
  to the mixture and marinate for 30 minutes.
- While the insects marinate, cook the rice according
  to the directions on the package.
- Drain the insects in a fine sieve over a bowl and reserve the marinade.
- Heat the oil in a wok and stir-fry the garlic, chili
  pepper, and ginger over high heat. Add the insects,
  mushrooms, snow peas, corn, and spring onions
  and stir-fry the mixture until everything is lightly
  browned. Add the cooked rice. Stir until heated
  through. Season to taste with the leftover marinade
  and salt, if necessary.

YIELD: SERVES 4

Chili con Carne. (Floris Scheplitz)

Chili con Carne

4 tablespoons olive oil
¼ pound (115 g)
    uncooked bacon,
    diced
2 garlic cloves, crushed
1 red chili pepper,
    seeded and thinly
    sliced
2 red onions, finely
    chopped
4 tablespoons tomato
    paste
1 tablespoon paprika
2 tablespoons all-
    purpose flour
½ cup (120 ml) red wine
1 (15-ounce [400 g]) can
    diced tomatoes
2 (15-ounce [400 g])
    cans kidney beans,
    drained and rinsed
1 tablespoon Tabasco
    sauce
4 cups (1 L) vegetable
    stock (homemade
    or prepared from a
    bouillon cube)
3½ ounces (100 g)
    mushrooms,
    quartered
1 red bell pepper,
    seeded and diced
1 green bell pepper,
    seeded and diced
⅔ cup (25 g) buffalo
    worms
¼ cup (10 g) mealworms
Salt and freshly ground
    black pepper
4 tablespoons chopped
    fresh parsley

- Heat 2 tablespoons of the olive oil in a large skillet over medium-high heat and fry the bacon, garlic, chili pepper, and onions until lightly browned. Add the tomato paste and paprika and fry for 1 minute longer. Sprinkle in the flour and continue to fry for another minute. Then add the red wine, followed by the tomatoes, beans, Tabasco sauce, and stock.
- Heat the other 2 tablespoons of olive oil in a separate skillet over medium-high heat and fry the mush-rooms and bell peppers for 5 minutes. Add this mixture to the chili and simmer over low heat for 15 minutes. Heat a lightly oiled nonstick pan over medium-high heat and fry the insects for about 1 minute before adding them to the chili. Adjust the seasonings with salt and black pepper and garnish with the parsley just before serving.

YIELD: SERVES 4

Tip

- Serve with tortilla chips, steamed rice, corn bread, or crusty French bread.

Chop Suey. (Floris Scheplitz)

Chop Suey

1½ teaspoons red curry
  paste
Juice of ½ lemon or lime
2 tablespoons mirin
  (sweetened rice wine)

12 grasshoppers, legs
  and wings removed
⅓ cup (15 g) mealworms

1 large carrot, cut into
  matchsticks
5 ounces (150 g)
  green beans, sliced
  diagonally
1 pound (450 g) dried
  soba noodles
4 tablespoons sunflower
  oil
8 ounces (250 g)
  mushrooms, cut into
  quarters or eighths
1 garlic clove, crushed
3 shallots, minced
1 yellow bell pepper,
  seeded and cut into
  strips
3 spring onions, sliced
  into rings
½ cup (120 ml) chicken
  stock (homemade
  or prepared from a
  bouillon cube)
2 tablespoons Asian fish
  sauce
2 tablespoons soy sauce
Splash of sherry
Salt (optional)
1 tablespoon arrowroot
  powder
4 tablespoons chopped
  fresh cilantro

- Mix the marinade ingredients in a bowl. Marinate the grasshoppers and mealworms in this mixture for 10 minutes.
- Bring a pan of salted water to a boil and blanch the carrots and green beans for 1 minute. Drain and rinse under cold running water.
- Drain the insects in a sieve over a bowl and reserve the marinade.
- Cook the soba noodles in plenty of boiling water, following the directions on the package. Drain well.
- Heat 2 tablespoons of the oil in a wok over high heat and stir-fry the mushrooms for 1 minute, until lightly browned. Transfer them to a bowl and set aside. Heat another 2 tablespoons of the oil in the wok and stir-fry the shallots, garlic, and insects until lightly browned. Add the carrots, beans, mushrooms, bell pepper, and spring onions. Stir-fry everything together, then add the stock, sherry, fish sauce, soy sauce, and reserved marinade. In a small bowl, mix the arrowroot with 3 tablespoons of cold water and add to the wok, to thicken the sauce a little. Season to taste with salt, if necessary, and cook for another 2 minutes.
- Divide the noodles among four soup plates or bowls and top with the stir-fried mixture. Garnish with the cilantro.

YIELD: SERVES 4

Jambalaya. (Floris Scheplitz)

Jambalaya

Equipment needed: deep fryer, with vegetable oil for frying

½ pound (220 g) boneless chicken, cut into small strips
1 tablespoon Cajun spice mix
Salt
4 cups (1 L) chicken stock (homemade or prepared from a bouillon cube)
1¼ cups (200 g) raw basmati rice
16 grasshoppers, legs and wings removed
3 tablespoons tempura batter mix (available at supermarket)
3½ tablespoons cold water
4 tablespoons olive oil
3½ to 4 ounces (100 g) chorizo, diced
½ cup (20 g) mealworms
2 garlic cloves, crushed
1 onion, minced
4 ribs celery, thinly sliced
1 red bell pepper, seeded and diced
4 spring onions, thinly sliced
Freshly ground black pepper
3 tablespoons chopped flat-leaf parsley

- Mix the chicken strips with 2 teaspoons of the Cajun spice and a pinch of salt.
- In a medium saucepan, boil the rice in the stock for 8 minutes, then pour off the excess liquid, cover, and let it sit for 15 minutes.
- Sprinkle the rest of the Cajun spice over the grasshoppers and drizzle them with 1 tablespoon of olive oil.
- Make a batter by combining the tempura mix, cold water, and a pinch of salt in a bowl. In the meantime, heat oil in the deep fryer to 350°F (180°C).
- Heat 3 tablespoons of the olive oil in a wok and stir-fry the chorizo for 2 minutes over high heat, until lightly browned and crispy. Remove the chorizo and set aside. Add and stir-fry the spiced chicken pieces in the same pan for 2 minutes, until done. Add the mealworms at the very end and fry them lightly. Remove the chicken and mealworms from the pan and set aside. Stir-fry the garlic and onions briefly; add the celery, bell pepper, and spring onions; and stir-fry. Mix in the chorizo, chicken, mealworms, and rice; season with salt and pepper to taste; and stir until warmed through.
- Using a fork, dip the grasshoppers in the batter, coating them well, and deep-fry them for a few minutes, until nicely browned.
- Serve in an attractive bowl, garnished with the fried grasshoppers and chopped parsley.

YIELD: SERVES 4

Insect Burgers. (Floris Scheplitz)

Insect Burgers

BURGERS

15 ounces (400 g) starchy
  potatoes, peeled and
  quartered
2 tablespoons sunflower oil
2 shallots, minced
1 ounce (30 g) carrot, very
  finely diced
1 cup (40 g) mealworms
1 ounce (30 g) canned corn
2 spring onions, sliced into
  very thin rings
4 large eggs
Salt and freshly ground
  white pepper
3 tablespoons all-purpose
  flour
4 to 8 tablespoons bread
  crumbs
2 tablespoons (30 g) unsalted butter, for frying

DRESSING

2 tablespoons rice wine
  vinegar
2 tablespoons chili sauce
1 tablespoon sesame oil
1 tablespoon sunflower oil
1 (½-inch [1 cm]) piece
  fresh ginger, peeled and
  grated
1 garlic clove, crushed

SAUCE

4 tablespoons oyster sauce
2 tablespoons Asian fish
  sauce
Juice of ½ lime or lemon
1 teaspoon brown sugar

4 ounces (100 g) mixed
  salad greens

- Boil the potatoes in plenty of salted water for 20 minutes until done. Drain, then steam them briefly, uncovered, on low heat.
- Mix the dressing ingredients in a jar, bottle, or shaker.
- Place the sauce ingredients in a small saucepan and bring to a boil. Boil for 1 minute.
- Heat the sunflower oil in a skillet over medium heat and fry the shallots gently, until soft but not browned. Add the carrot and mealworms and cook gently for 2 minutes until the carrots are soft; remove from the heat.
- Mash the potatoes with a potato ricer or masher. Add the mealworm mixture, corn, and spring onions. Stir in two eggs and season with salt and pepper.
- Shape the mixture into four flat patties, using a bit of flour to prevent sticking. Chill the patties well in the refrigerator. Break the remaining two eggs into a soup plate and whisk them well with a fork. Sprinkle the bread crumbs onto another platter. Dip the patties in the egg, then coat them with bread crumbs. Allow the patties to firm up for 15 minutes, then dip them again in the egg and bread crumbs.
- Melt the butter in a skillet and fry the patties over medium heat, until crisp on the outside and warmed through, about 8 minutes.
- Toss the salad greens with the dressing and serve with the insect patties. Spoon the sauce onto each plate, next to the patties.

YIELD: 4 BURGERS

Vols-au-vent. (Floris Scheplitz)

Vols-au-vent

Equipment needed:
baking sheet

4 puff pastry vol-au-vent cases or patty shells

RAGOUT

⅔ cup (25 g) mealworms
¼ cup (10 g) buffalo worms
4 tablespoons (50 g) unsalted butter
1 onion, finely chopped
Scant ½ cup (60 g) all-purpose flour
1⅔ cups (400 ml) vegetable stock (homemade or prepared from a bouillon cube)
3½ tablespoons heavy cream
Salt and freshly ground black pepper
1 large egg yolk
1 tablespoon chopped fresh parsley

Preheat the oven to 250°F (120°C).

- Cut off the tops of the pastry cases and put the cases and their tops on a baking sheet. Bake for 30 minutes.
- Toast the insects for 2 minutes in a lightly oiled nonstick pan over medium-high heat until lightly browned; set aside.
- Prepare the ragout: Melt the butter in a saucepan and fry the onion over medium heat until soft but not browned. Whisk in the flour. Cook the roux over low heat, whisking, for 2 minutes. Add half of the stock and whisk until smooth. Add the rest of the stock and simmer the ragout gently for 10 minutes, stirring regularly to prevent scorching. Add the insects and season to taste with salt and pepper.
- In a bowl, mix the egg yolk and cream and add this, along with the parsley, to the warm ragout. Fill the warm pastry cases and serve immediately.

YIELD: SERVES 4

Tip

- This ragout can also be served with steamed rice or pasta.

Quiche. (Floris Scheplitz)

Quiche

Equipment needed:
10-inch (25 cm) quiche pan

DOUGH

1 cup plus 2 tablespoons (150 g) all-purpose flour
1 teaspoon active dry yeast
1 teaspoon salt
1 teaspoon superfine sugar
3½ tablespoons butter, plus more for greasing the pan
1 large egg
4 tablespoons milk

FILLING

⅔ cup (25 g) mealworms
4 tablespoons olive oil
2 onions, chopped
½ leek, sliced finely and well rinsed
1 red bell pepper, seeded and diced
1 garlic clove, crushed
1 teaspoon curry powder
1 scant cup (220 ml) heavy cream
4 large eggs
1 tablespoon cornstarch
½ cup (50 g) grated aged cheese
Salt and freshly ground black pepper

Preheat the oven to 350°F (180°C).

- In a bowl, mix the dough ingredients and knead until smooth. Shape the dough into a ball, cover, and allow to rise in a warm place for 15 minutes.
- Prepare the filling: Chop the mealworms into small pieces. Heat the oil in a skillet over medium-high heat and fry the mealworms, onions, leek, bell pepper, and garlic for about 2 minutes, without browning. Add the curry powder and fry for another minute. Allow to cool.
- Mix the cream, eggs, cornstarch, and cheese in a bowl and season with salt and pepper.
- Roll out the dough to a thin circle, using a bit of flour to prevent sticking. Butter the quiche pan and line it with the dough. Spread the insect mixture out evenly on the dough and pour the cream mixture on top. Bake the quiche for 40 minutes in the middle of the oven until set and lightly browned.

YIELD: SERVES 8

Tip

- If you prefer the mealworms to be clearly visible in the pie, leave them whole.

DANIELLA MARTIN, EDIBLE-INSECT ADVOCATE

"Valuable, Abundant, and Available to Everybody"

Girl Meets Bug is the title of a video series created by Daniella Martin to show how easy, eco-friendly, and tasty eating insects can be. She has also created a Web site called Girl Meets Bug, with information about entomophagy, sources of edible insects, recipes, and links, as well as demonstration videos.

AZTEC AND MAYA

In Vermont, Daniella Martin studied cultural/medical anthropology at a small college. She was most interested in pre-Columbian nutrition and medicine. Her question in particular was: What vestiges of early Aztec/Mayan life still existed? So she went to the Yucatán in Mexico and conducted research with local Maya on the various types of food that they ate as well as the various types of botanicals they utilized. During her study she read a great deal about how—despite the scarcity of large game in seventeenth-century Mexico—the Aztecs still had plenty of protein. "I remember reading about an argument between different schools of thought on Aztec cannibalism. One side thought that the Aztecs made a conscious spiritual choice of communicating with their gods via consuming human bodies (the bodies of their victims became divine). The other side suggested that they practiced cannibalism because they did not have enough protein: lacking large game, and having only small dogs and poultry, they were biologically driven to eat each other. That was really a big statement. Then a bunch of specialists and anthropologists who studied this particular region came back and said, 'No, you've got it all wrong'. The Aztecs were not suffering from protein deprivation. They had a broad variety of protein sources. They ate everything that swims and crawls, including—especially—insects."

Daniella Martin. (Daniella Martin)

TEXTBOOK COME ALIVE

There she was, a very young anthropologist sitting in the college library and reading about cannibalism and edible insects. It was too fascinating for words. That began in 1999; in 2000, she lived in Yucatán for eight months, studying firsthand what vestiges had survived. "I remember that I was in Oaxaca and a woman was selling *chapulines*: roasted, chili and lime–spiced grasshoppers. For me, this was a textbook come alive. It was now right before my eyes: first reading about this argument about cannibalism and entomophagy, and now it was real and definitely not dead and still existing. And then somebody brought me a bag of grasshoppers and I sat down at the café, eating them and taking pictures of them. My table was then surrounded by Mayan kids coming from the street into the café to eat my *chapulines* right off my table. They did not even ask; they just started eating them and smiling at me. As a burgeoning anthropologist, this was a cultural immersion experience at its best."

CHILDREN

Martin is all enthusiasm: "I love children, and they became an important factor in all my studies. It started with my Mayan informants who did not speak Spanish. Everything had to be translated through their kids, from Mayan to Spanish and then into English. So children were a vessel for my research. The children ate insects, not because it was good for the environment, or because of the rich protein contents and the few greenhouse gases. No, they enjoyed the taste of the insects. That, for me, was a kind of flash. So it started from an anthropological perspective."

OVERLOOKED

"Insects as food are seen very differently in the United States," Martin continues, "but they could be a popular snack food and a good source of protein in the United States, too. My interest is in things that people overlook and ignore, despite being valuable, abundant, and available to everybody. In the United States, nobody thought about eating insects, even though it was such a logical idea."

WAX MOTH LARVAE

In 2008, she read in *Time* about a workshop on edible insects, sponsored by the Food and Agriculture Organization of the United Nations (FAO), held in Chiang Mai, Thailand, and the ecological potential of edible insects.[1] The article mentioned two Americans, David Gracer and David Gordon, both advocates of consuming insects. "I realized that eating insects could really happen, and that there are good solid reasons for it, capable of improving the world. Six months later I put up a blog. A short time after that, I called David Gracer and went to his house, and he served me insects from all over the world. I phoned David Gordon [the 'Bug Chef' who wrote *The Eat-A-Bug Cookbook*] to discuss what kind of bugs to start with. In late 2008, I cooked my first bug meal, with wax moth larvae. I did a lot of research on which insects are safe to eat, how they are raised, what one feeds the crickets and the mealworms. The answers were not particularly satisfactory, until I called a wax moth producer, who said, 'We feed them bran and honey.' That was good enough for me, so I ordered the first batch of 250 live wax moth larvae. I sorted them out from the wood shavings, froze them, sautéed them with onions and olive oil, and made tacos out of them. Tacos, because I lived in Mexico; this food is very normal for me. I was already familiar with the *gusanos* [caterpillars], such as the agave worm, which Mexicans have been eating for centuries. I remember taking my first bite into the taco with the wax moth larvae and thinking, 'There is nothing wrong with this, it is great!' This was objectively tasty. It was like a wonder to me. So I spent more and more time on this."

Her boyfriend bought her the book *Ecological Implications of Mini-livestock* as a birthday present, and that really set her off. "I started to think, 'What can I do, to add to what has already been done?'"

THE RACHAEL RAY OF BUGS

The idea came from a job she had held the previous year in Silicon Valley. Martin worked on an online game show, where people from all over the world played against one another. She was the host, and on the air essentially every day, with a lot of camera time. It was always a dream to become a travel show host. "My idea was: 'Some day, I can be a travel writer, or host a travel show, and I can work the edible insects into my other career.'"

Martin continues: "I have never seen a cooking show with insects, and many of the hosts of exotic cooking shows—or the authors of weird-food cookbooks—are men. The only exceptions are Dr. Julieta

1. Brian Walsh, "Eating Bugs," *Time*, May 29, 2008, http://www.time.com/time/magazine/article/0,9171,1810336,00.html.

Ramos-Elorduy [author of *Creepy Crawly Cuisine: The Gourmet Guide to Edible Insects*, together with Peter Menzel]; Faith D'Aluisio, Peter Menzel's wife [*Man Eating Bugs: The Art and Science of Eating Insects* by Peter Menzel and Faith D'Aluisio]; and Dr. Florence Dunkel. It would be helpful for the public to see a woman, because women are associated with the home and with food. Rachael Ray was a big deal at the time, a great example. She was revolutionary. She did a travel show, a Food Network show called *$40 a Day*, and she came out with her own magazine, *Every Day with Rachel Ray*. I thought, 'Maybe I can become the Rachael Ray of bugs.' That was a sort of my goal, and I made two videos; they were pretty well received. I got a lot of attention from them, and I actually moved to Los Angeles to work with a producer on a video project. However, he stopped, and producing videos on my own was just not on. But I started my own Web site, Girl Meets Bug, which is receiving a lot of interest."

UNITED STATES CENTER OF ENTOMOPHAGY

Although it may seem that entomophagy is not well developed in the United States, this is not exactly true. Where is the center of entomophagy in the United States? Daniella quickly concludes, "Oh, San Francisco. Not only because of me, but also because of Monica Martinez, who, at the 2011 San Francisco Street Food Festival, launched the nation's first edible insect food cart, Don Bugito. The new Chapul company also has its headquarters in San Francisco: Chapul Bars are all-natural bars with protein from crickets, inspired by native techniques used for centuries in the American Southwest and Mexico. The company is based in Salt Lake City and San Francisco. And then there is yet another company called Infood. San Francisco is definitely a hub. The other hub is probably New York City; they have the most restaurants that serve insects, but still not very many."

INSECTS ARE *SO* GOOD!

Her plans for the future are to write a book to reach people who wouldn't ordinarily be interested in entomophagy. "I want to write it in such a way that it goes beyond the entomophagy aspects, and includes the type of stories that people want to read, for a broad audience. Lectures on entomophagy were the main thing I was doing, but it takes a lot of energy; now I'm concentrating on writing the book."

Her favorite dish is wax moth larvae: "They are great. You can roast them in the oven. They are excellent in almost every recipe that I make. The taste is like mushrooms and nuts at the same time. Everybody I present them to finds them fantastic. One of my favorite recipes for them remains sautéing, as I did the very first time I ate them. I first sauté some chopped onions in butter, and then add the wax moth larvae just as I would any other meat, and a little salt. When the larvae become slightly golden and translucent around the edges, they're done. They're delicious in tacos, with a little lettuce, avocado, and salsa. So good! *So* good."[2]

2. For the tacos, see *Girl Meets Bug*, episode 1, "Waxworm Tacos," http://www. youtube.com/watch?v=fA_rBNeVtzo.

ROBÈRT VAN BECKHOVEN, PASTRY CHEF

"Bonbon *Sauterelle*"

Bij Robèrt (Dutch for "Chez Robèrt") is the culinary studio of master pâtissier (pastry chef) Robèrt van Beckhoven. It is a trendy space with a large, open kitchen full of stainless steel: bright, clean, and inviting. He welcomes his guests here, and teaches workshops. Downstairs is the production line, where his employees make the most delicious cakes, and where the framework for what will be a famous soccer player's wedding cake towers high.

FROM PASTRY . . .

Robèrt van Beckhoven began as a bread-and-pastry baker, has won every prize awarded in his profession, and is, since last year, one of the two master *pâtissiers* of the Netherlands. At important events, tennis tournaments, and large Joop van den Ende Stage Entertainment productions, van Beckhoven and his employees are the house *pâtissiers*.

It is remarkable that someone with his credentials would make the move onto the experimental path. Or isn't it? "Actually, I have always been involved in innovative things. Out-of-the-box thinking is very important to me. Above all, I want the products that I make to be inspiring. If people ask me to develop a new product,

Robèrt van Beckhoven. (Lotte Stekelenburg)

it always means small amounts. So there I stand, making Brussels sprouts ice cream for thirty people on my own, because the quantities are too small for the production line."

. . . TO INSECTS

The step from experiments like blowing bubbles with syrup to products with insects is, according to Van Beckhoven, not all that big. When two students from the "Food Design and Innovation" university program asked him to make a granola bar with insects, he went right to work. "To me, anything new is a challenge," was his reaction. The granola bar won an award.

While visiting a food fair with food of the future as its theme, he saw dishes made with insects. "Interesting, but they looked just terrible. A gray mush. Immediately, I remarked that it would also be possible to make it pretty and appealing." Not long thereafter, he presented tidbits to people attending the "Food Inspiration" conference. "In the morning there were bites that no one knew contained insects; afternoon visitors were told about the ingredients; and the bites in the evening sported an insect garnish on top. You have to begin imperceptibly, with something that looks familiar, pretty, and appealing. If it also tastes good, most people don't think it's so bad when it turns out to contain insects. After that, you can make it more and more visible."

MOTIVE

"An insect is simple. What's a little insect? Do you know how many insects you are already ingesting by eating your normal food? There's nothing wrong with that. It's not so remarkable, but we just don't know about it. And let's face it: we eat snails and frogs' legs, and mussels, so it's not about that. People have to learn, in these times of socially responsible business practices, that producing meat is much more expensive than producing insects. And that farming insects is much less of a burden on the environment." This is also Van Beckhoven's biggest motive for promoting insects as food.

One of Robèrt van Beckhoven's creative extravagances: petit four with grasshopper. (Hans Wolkers)

How does he go about that? In the cooking classes he teaches at his studio, he likes to challenge people. He asks them to snort cocoa; at some point he also shows them a box of mealworms. He asks if people dare to taste them, and he whips something simple up right then and there. Something with chocolate, which is familiar. And that is the moment when Van Beckhoven can tell people what, to him, is the important information about insects: that they are healthy and full of protein. He sees it as his task to get women who scream at the sight of insects to eat them. "You have to challenge people, but also allow them to adjust. It shouldn't be too creepy in the beginning. Build it up. Make it appealing. And especially: explain its importance.

"You have to be careful when preparing insects. Visitors who see a couple of mealworms lying on a cutting board will think that they are pests, that your kitchen is unhygienic. Let them see the package, not the individual insects."

EATING INSECTS: HOW, THEN?

What other products does Van Beckhoven think will win people over? "A quiche is easy and tastes good, and the insects in it don't have to be identifiable. That's also true for cookies, of course: you can easily add ground mealworms. You can also cast insects into a lollipop, which looks fabulous. And all combinations with chocolate go over well. But then the chocolate has to be organic. And consider the name you use: 'bonbon *sauterelle*,' with the French word for 'grasshopper,' sounds tastier than 'grasshopper bonbon.'"

Roasted grasshoppers at the Klong Toey market in Bangkok, Thailand. (Arnold van Huis)

Cochineal from Peru

Advertisements for strawberries bedazzle us with their bright red colors. Sometimes, however, these much-loved summer berries are just not red enough. Producers of strawberry yogurt, for example, regularly use an insect to color their product more convincingly. The scale insect cochineal (*Dactylopius coccus*) produces a red dye called carmine when it is crushed. Scale insects are sucking insects that are related to aphids and feed on plant juices. The cochineal lives exclusively on *Opuntia* cactuses, known as paddle cactuses because they grow as rounded overlapping paddles armed with spines. Colonies of cochineals grow on these paddles. To harvest them, the insects have to be scraped off the cactus plants. More than 100,000 insects are needed to make 2.2 pounds (1 kg) of cochineal extract. A 2.5-acre

Cupcake with icing colored with Natural Red 4, or E120. (Lotte Stekelenburg)

(1 hectare) field of cactuses can produce 660 to 880 pounds (300 to 400 kg) of these insects. Eighty percent of cochineal produced for use around the world comes from Peru, supplemented by cochineal production on the Canary Islands and in China.

Cochineal extract has been used for centuries in aquarelle and oil paints and as a fabric dye. It is also found in cosmetics such as lipstick

Cochineal scale insects on a paddle cactus. (Zyance)

Close-up of cochineal insects on paddle cactus. (Zyance)

and nail polish. Cochineal is ideal for achieving a deep red color, and it is also used for coloring candies, alcoholic drinks, and surimi (imitation crab). It is known as the food additive Natural Red 4 or E120, carmine (its pure form), or carminic acid. Our food, therefore, very frequently contains a natural insect dye.

Maggot Cheese in Sardinia

The Netherlands has its Limburger cheese; France, its Roquefort; and Sardinia, its *casu marzu* (literally, "rotten cheese"). This cheese is made from sheep's milk, and it is left to ripen for so long that it starts to rot. The rotting process attracts cheese flies that then lay their eggs on the cheese. The fly larvae eat their way into the cheese, digesting fats along the way and making the hard cheese soft and runny on the inside. The cheese is eaten, larvae and all, just as Roquefort is eaten with its mold. The larvae are about ⅓ inch (8 mm) long and colorless. The cheese is at its best when fluids start leaking out of the crust. The taste is quite sharp. *Casu marzu* is banned in the United States. Although

Maggot cheese. (Shardon)

a new European Union regulation allows its local sale as a "traditional food," this Sardinian specialty is now otherwise prohibited because it is unlawful to sell a spoiled product. But are fly larvae really any different from the bacteria and fungi in Limburger cheese or Roquefort or in the Swedish fermented herring, *surströmming*?

Palm Beetles in the Tropics

In the tropics, palm beetle larvae are considered a delicacy. Palm beetles belong to the weevil family: they are beetles with a long snout that allows them to bore into plants. The species of palm weevils eaten in Africa, Asia, and Latin America vary, but are all closely related.

When a palm tree (coconut, oil or sago palm) is knocked or chopped down, the beetles are more likely to lay their eggs on them. The emerging larvae feed on the contents of the tree trunks.

Palm beetle larvae. (Luigi Barraco)

In Africa and Latin America, women have a technique for estimating when the larvae are large enough to be harvested and eaten. They press their ear against the tree trunk and listen to the munching sounds of the larvae. The quality of the sound indicates the approximate size of the larvae. When the women give the go-ahead, men get started chopping open the tree trunks to harvest the larvae, which can then be sold at the market.

After the larvae have been rinsed off, they are panfried. No extra oil is needed because their bodies already contain enough fat, but people often add onions, pepper, and salt. Another cooking technique used is to skewer the larvae and barbecue them.

In Nigeria, children are told not to eat palm weevil larvae because it will get them drunk—because palm is also used to make palm wine. It is, of course, nonsense that the larvae could cause drunkenness, but many parents use this tall tale to prevent their children from chopping down too many palm trees essential for the production of palm oil and palm wine.

Palm beetles at the market in Yaoundé, the capital of Cameroon. (Arnold van Huis)

Dragonfly Larvae in China

Dragonflies are usually found near water. Large colonies live around irrigated rice fields in Asia. The use of pesticides against rice pests, however, has caused a rapid decrease in their numbers.

A dragonfly's head is composed mainly of large, bulging eyes which they need to catch other insects on the wing. For that, they also have to be expert aviators: they are able to perform agile aerobatics. Their excellent eyes keep track of everything, and although these insects are difficult to catch because of their flight maneuvers, children in China are excellent dragonfly catchers. The children spread a sticky sap from a certain tree onto the end of a long bamboo stick and then reach out and hold the sticky end over a resting dragonfly. When they touch it to the dragonfly, it becomes glued to the stick and can't get away.

After the sticky sap is cleaned off the dragonflies with a bit of vegetable oil and their wings are removed, they are ready to cook. You can stew them for 5 to 10 minutes in coconut milk with ginger, garlic, onions, and chili peppers. You can also barbecue them.

People like the dragonfly larvae even better than the adult insects. Dragonfly females dip their abdomen into the water and lay their eggs

Diner eating dragonfly larvae in a restaurant in Dali, China. (Marcel Dicke)

Dish made with dragonfly larvae and deep-fried peppermint leaves, served at a restaurant in Dali, China. (Marcel Dicke)

on underwater plants. Larvae that hatch from these eggs are often scooped out of the water along with other insects and eaten. In the Chinese city of Dali, which is located at the edge of a lake, dragonfly larvae are a popular food. They are often kept in containers in restaurants, and you can point out which ones you would like to eat, as is sometimes done with seafood in restaurants in the West. The chef will then fry a portion of about forty dragonfly larvae with some peppermint leaves—served with rice, this makes a festive meal.

Live dragonfly larvae at the entrance to a restaurant in Dali, China. (Marcel Dicke)

FIVE FESTIVE DISHES

Chebugschichi. (Floris Scheplitz)

Chebugschichi

SAUSAGES

⅔ cup (25 g) mealworms
1 pound (450 g) ground
  beef
1 onion, finely chopped
1 large egg
2 tablespoons bread
  crumbs
½ teaspoon paprika
¼ teaspoon chili powder
Salt and freshly ground
  black pepper
6 tablespoons olive oil

BARBECUE SAUCE

1 onion, minced
1 tablespoon vegetable oil
4 tablespoons ketchup
1 slice canned pineapple,
  drained and finely
  chopped

GARLIC SAUCE

4 tablespoons mayonnaise
2 tablespoons plain yogurt
1 garlic clove, crushed
¼ teaspoon prepared
  mustard

4 hamburger buns
Handful of green curly-
  leaf lettuce
1 red onion, thinly sliced
  into rings
¼ cucumber, sliced
1 red bell pepper, seeded
  and cut into strips

The name of this dish is derived from a popular dish from Istria, in Croatia, called *cevapcici*; the hamburger has been replaced with sausages.

- Prepare the sausages: In a bowl, knead together the ground beef, mealworms, onion, egg, bread crumbs, paprika, and chili powder. Season with salt and pepper.
- Shape this mixture into twenty small sausages a little more than 2 inches (5 cm) long.
- Heat the olive oil in a skillet over medium-high heat and fry the sausages for about 5 minutes, until browned and cooked through.
- Prepare the barbecue sauce: Fry the onion in the oil for a few minutes over medium-high heat, add the ketchup and pineapple, and simmer for 2 minutes over low heat.
- Mix all the garlic sauce ingredients in a bowl.
- Cut open the hamburger buns and toast them lightly in a grill pan or dry skillet.
- Garnish each warm bun with the lettuce, five sausages, and some red onion rings, cucumber slices, bell pepper strips, and the sauces.

YIELD: 4 SANDWICHES

Tip

- If you want to work with fresh mealworms, use ½ pound (225 g) of ground meat and ½ pound of fresh mealworms. Rinse the mealworms and blanch them for 1 minute in boiling water. Rinse with cold water, drain well, and chop them briefly in a food processor. Mix them into the other ingredients as described.

Hopper Kebabs. (Floris Scheplitz)

Hopper Kebabs

Equipment needed:
8 skewers

MARINADE

1 garlic clove, crushed
1 teaspoon shrimp *sambal* (shrimp-chili paste)
2 tablespoons sesame oil
2 tablespoons stem ginger syrup
6 tablespoons soy sauce
Juice of ½ lime

KEBABS

24 grasshoppers, legs and wings removed
4 fresh baby corn cobs, halved
½ zucchini, seeded and cut into chunks
8 button mushrooms, halved
4 tablespoons peanut, safflower, or canola oil
8 cherry tomatoes

- Heat the marinade ingredients in a saucepan over medium-high heat and simmer for 1 minute. Allow to cool, reserving some of the marinade for serving.
- Prepare the kebabs: In a small bowl, toss the grasshoppers with 3 tablespoons of the marinade.
- Cook the baby corn cobs in salted water for 10 minutes, until crisp-tender. Rinse them under running cold water.
- Place the zucchini, mushrooms, and corn cobs in a bowl and drizzle them with 3 tablespoons of the oil. Heat a grill pan until very hot and grill the vegetables. Transfer them to a platter and allow to cool.
- Heat the rest of the oil in a skillet over medium-high heat and fry the insects for 2 minutes.
- Alternate the grilled vegetables, cherry tomatoes, and insects as you thread them onto the skewers.
- Place the skewers on a platter and brush with some of the reserved marinade.

YIELD: SERVES 4 (2 KEBABS PER SERVING)

Tips

- If using canned baby corn cobs, they do not have to be precooked before grilling.
- If serving the kebabs later, reheat them in a preheated 250°F (120°C) oven or serve them cold.

Pizza. (Floris Scheplitz)

Pizza

Equipment needed:
baking sheet

DOUGH

2¼ cups (300 g) all-purpose
    flour, plus more for
    rolling out
2 teaspoons active dry yeast
1 teaspoon salt
1 teaspoon superfine sugar
2 tablespoons olive oil
¾ cup (180 ml) lukewarm
    water

TOMATO SAUCE

2 tablespoons olive oil
1 small onion, finely
    chopped
½ red chili pepper, seeded
    and finely chopped
1 garlic clove, crushed
4 tablespoons tomato paste
1 (15-ounce [400 g]) can
    diced tomatoes
1 bay leaf
1 sprig fresh rosemary
Salt and freshly ground
    black pepper

TOPPING

2 tablespoons olive oil
½ pound (225 g) mushrooms
½ zucchini, diced
12 grasshoppers, legs and
    wings removed
½ cup (20 g) mealworms
1 tablespoon capers,
    drained
20 pitted black olives
4 ounces (100 g) aged
    cheese, grated

Preheat the oven to 450°F (240°C).

- In a bowl, mix the dough ingredients and knead until smooth and elastic. Divide the dough in half, shape into two balls, cover, and allow to rise in a warm place for 15 minutes.
- Prepare the tomato sauce: Heat the olive oil in a saucepan and fry the onion, chili pepper, and garlic over medium-high heat until softened. Stir in the tomato paste and fry for 2 more minutes. Add the diced tomatoes, bay leaf, and rosemary and simmer for 30 minutes. Add salt and pepper to taste and re-move the bay leaf and rosemary.
- Prepare the topping: Heat the olive oil in a skillet over medium-high heat and fry the mushrooms and zucchini for about 3 minutes.
- Roll out the balls of dough to make two round, thin pizza crusts, dusting with flour to prevent sticking. Place the crusts on a greased baking sheet.
- Spread about five large spoonfuls of tomato sauce on each crust. Distribute the mushrooms, zucchini, grasshoppers, mealworms, capers, and olives over the sauce. Sprinkle with the grated cheese.
- Bake the pizzas in the middle of the oven for about 15 minutes until done.

YIELD: 2 PIZZAS

Bugitos. (Floris Scheplitz)

Bugitos

1 ripe avocado
1 garlic clove, crushed
1 teaspoon Dijon
   mustard
1 tablespoon
   mayonnaise
Juice of ¼ lemon
Pinch of salt
Freshly ground white
   pepper

FILLING

1 tablespoon lime oil
20 grasshoppers, legs
   and wings removed
Salt
¼ teaspoon chili
   powder

4 flour tortillas

¼ head iceberg lettuce,
   shredded
4 tomatoes, seeded
   and cut into strips
2 spring onions, sliced
   into thin rings
4 tablespoons chili
   sauce
½ cup (120 ml) sour
   cream

- Prepare the guacamole: Cut the avocado in half, remove the pit, peel, and dice. Add the garlic, mustard, mayonnaise, and lemon juice and puree with a stick blender or in a food processor. Season to taste with salt and pepper.
- Prepare the filling: Heat the lime oil in a wok over medium-high heat and fry the grasshoppers for 2 minutes, until crispy. Transfer them to a platter and sprinkle with a little salt and the chili powder.
- In a dry skillet, heat the tortillas on both sides. Divide the guacamole, lettuce, tomato, spring onions, grasshoppers, chili sauce, and sour cream evenly among the four warm tortillas. Roll them up, cut them into bite-sized pieces, and hold each piece together with a toothpick.

YIELD: SERVES 4

Crêpes. (Floris Scheplitz)

Crêpes

BATTER

⅔ cup (100 g) raisins
1⅔ cups (400 ml) milk
2 teaspoons active dry
    yeast
⅓ cup (50 g) superfine
    sugar
Scant 2 tablespoons
    (10 g) unsalted butter
2 cups (250 g) all-
    purpose flour
½ cup plus 1
    tablespoon (50 g)
    uncooked quick oats
Pinch of salt
2 large eggs
½ cup (20 g) buffalo
    worms

Sunflower oil, for frying

CINNAMON SUGAR (OPTIONAL)

4 tablespoons
    powdered sugar
2 teaspoons ground
    cinnamon

Golden syrup (optional)

- Prepare the batter: Wash the raisins and soak them in lukewarm water for 15 minutes.
- Over low heat, in a saucepan, heat a scant ½ cup (100 ml) of the milk until lukewarm. Stir in the yeast and superfine sugar. Remove the pan from the heat, cover with a damp tea towel, and set it to rest in a warm place.
- In a small saucepan, heat the remaining milk and butter until the butter has melted. Cool to lukewarm.
- Combine the flour, oats, and salt in a medium bowl; make a well and pour in the yeast mixture. Add about half (a little more than ½ cup [about 135 ml]) of the butter mixture. Using a whisk, stir from the center outward, to incorporate the liquid. When the mixture is smooth and all lumps are gone, add the rest of the butter mixture and the eggs, and beat for a few minutes to make a nice batter. Cover it again with a damp tea towel and set it to rest in a warm place for 15 minutes.
- Drain the raisins well and incorporate the raisins and the buffalo worms into the batter with a spatula. Cover and allow to rise in a warm place for 30 minutes.
- Heat a little oil in a skillet over medium-high heat and ladle a generous amount of batter into the center. Lift and rotate the pan so that the batter covers the bottom. Shake now and again to loosen the crêpe and allow it to brown nicely. Flip or turn the crêpe and brown the other side.
- While frying the crêpes, mix the powdered sugar and cinnamon, if using, in a small bowl.
- Transfer the crêpes to a warm plate and serve with cinnamon sugar or golden syrup.

YIELD: SERVES 8

RENÉ REDZEPI, TOP CHEF

"An Exploration of Deliciousness"

For novel developments in haute cuisine, the capital of Denmark is the place to be. René Redzepi is a Danish chef, and co-owner of the two–Michelin star restaurant Noma in the Christianshavn neighborhood of Copenhagen. His was voted the best restaurant in the world in the San Pellegrino Awards three times in a row: in 2010, 2011, and 2012. The name Noma is a conflation of the Danish words for "Nordic" and "food." Redzepi is famous for his major contribution to the invention and refinement of a new Nordic cuisine, and of food characterized by inventiveness and clean flavors. He also founded the Nordic Food Lab, located on a boat just outside the restaurant, for the gastronomic exploration of Nordic cuisine.

THE *WOW* MOMENT

René Redzepi's interest in insects as food originates from an event that Noma organizes each year: the MAD symposium. This meeting brings

René Redzepi. (Ditte Isager)

together in Copenhagen the food world's brightest and most inventive minds—from chefs to farmers, from journalists to politicians. At the first symposium, in 2011, the theme was vegetation. A chef from Brazil, Alex Atala,[3] was there to speak. He said, "We are all talking about vegetation and vegetables as a delicious alternative to food, and what is good for us and what is good for the planet. But we never talk about insects, and in my part of the world, insects have always been a part of the diet for many." Redzepi comments: "So, okay, we listened. But then Atala served a snack made of a fat Amazonian ant, and I popped it in my mouth and chewed once and the flavor exploded—a mixture of lemongrass and ginger. For me this was one of those *wow* moments. This tiny creature had so much flavor, I did not understand it. It was as if I was drinking the juice of lemongrass, so to speak."

ANTS, LEMON, AND CORIANDER

"Then I thought, 'What about the ants that we have here in Denmark?' Are you telling me that the only edible ones are those from the Amazon? Do we potentially have such delicacies in Denmark, such intense flavors living right next door in the wild? And if so, can we harvest those? From that moment on, the exploration started. It was never a way of feeding the world in a sustainable fashion; to us it has been an exploration of deliciousness.

"We use two varieties of ants," the chef continues, "and one is intensely lemony. Of course, you could just use a lemon, I guess, but lemon trees do not grow here. This ant is more than that. It is a different nuance, a different flavor, and we don't have such lemony flavors normally. In the type of restaurant that we run, we try to be as local as possible. We use the farms that surround us. Although they do not have lemons, the lemon was suddenly there, through an ant. We harvest it sustainably; we learned how to do it. We make a paste of it, so we can flavor foodstuffs." Redzepi elaborates with excitement. "And then there is this other Danish ant that tastes like coriander, which is exotic to our part of the world. Coriander is a spice from Thai or Mexican cuisine. But here in Denmark it is not a plant, but a tiny creature. It was an amazing thing. There are beets that you cook all winter long, but suddenly you now have coriander to add to that equation; it's a magical tool for cooking."

3. Atala is the world-renowned chef-owner of D.O.M. and Dalva e Dito restaurants in São Paulo, Brazil.

HONEY

"In Europe we do not have a history of cooking insects. Honey is the only thing, and totally accepted, although processed through an insect. Maybe people do not realize anymore that it comes from an insect. I am sure that if you would discover that, say, the vomit of some kind of larva would be delicious, people would have a hard time accepting it. If there was strange-looking stuff in front of you, and people would say: 'Eat this, it is the vomit of a larva, from a disgusting insect', people would say 'No, no, no.' That was also one of the things we truly understood, that insects are a simple, yet so complex barrier that we have. But to me, the flavor of the ant was too good to miss, so we started to incorporate it into our cooking."

Redzepi does not utilize insects as a meat substitute. He feels that the driving force, especially in the Western world where most people can eat well, has to be deliciousness. It is simple: if it does not taste good, you will not be able to persuade anyone. "We found ways of making the two or three insect ingredients taste good—the ones we have experimented with so far, two types of ants and crickets. We made them tasty, very delicious. We ferment house crickets, turn them into a paste or liquid that is somewhere in between a fish sauce, a soy sauce, and a Mexican *mole*; really interesting."

INSECT CAVIAR

Redzepi tasted insects many times. He traveled to Mexico: "I tried ant eggs [larvae, also called 'insect caviar']. In Mexico, the dish is called *escamoles*. Stewed together with spices and tomatoes, it was a magnificent flavor and very tasty. The question was: Can we get these ants? Do our ants have the same taste, and how do you collect them? In Mexico you eat them in the finest of restaurants. I was at a very fine dinner with all sorts of people in Mexico this January; these ants with some kind of vegetable were the star of the menu. *Escamoles* is like white truffles. So these are some examples of insects—besides honey—as delicious food."

LOCAL PRODUCTS

Regarding the origin of the insects that Noma uses, Redzepi's vision is clear. "We explore the space we live in. Distilling the flavor of this region is what we are trying to do. It involves organisms in the ocean, in the forest, on shorelines, on the fields; crops that farmers can grow well here, or domesticated and wild animals, and also what is in between: the insects. But insects are a new addition to the history of Scandinavian cooking. Because of the very different seasons, insects are not always available. You can extend them, if you ferment and process them. But if you need a full meal of crickets, you need a whole bowl, maybe even two times a salad bowl. They take a lot of space, yet they weigh nothing, and in crickets there is not much flavor. And if you cannot add flavor, I do not believe people are going to eat them." Is this similar to meat that needs seasoning to make it tasty? Redzepi explains the difference. "Meat is full of umami; it is juicy. A cricket, just fried, is crunchy, and there is not a lot of flavor. Some are perfectly good as bread crumbs: they can be dried and ground to a powder, and then you can roll carrots in it, and fry them in butter. The result is browned and tasty. But right now, to eat insects as a meal, I believe we need a lot more experimenting, a lot more iconic dishes that taste really good, for people to be persuaded."

PRIMITIVE

One of the biggest hurdles to overcome in the Western world is that cultures that do eat insects are considered primitive. "This is a sort of racial thing; most Westerners have it subconsciously, and they won't admit it. They simply think, 'If they were as clever as we are, they would not eat insects, either.' I do not understand what happens, why it is like that, why people are so afraid of eating insects, why people are so conservative in terms of their food, but people just are." Redzepi sees that the situation is changing. "We have a dish with ants on the menu. All the guests just now ate it for lunch." Yet Redzepi sees that, in a globalizing world, people more easily adapt to other types of food. "People are more open, more curious, more open to being surprised. They acknowledge that these flavor experiences are a lovely part of life."

FLAVOR

At Noma, the guests know what they eat. "Everybody knows that we are experimenting. We do occasionally—maybe once or twice a week— have guests who do not wish to eat anything from the insect world. And then of course we ask, 'including honey?' Then they say, 'No, no, no, honey is fine,' but they have an aversion to eating insects. Even though we do not do it like this [he points to a picture of a muffin topped with a grasshopper]. Right now we are disguising it." Redzepi's interest is the flavor, and the ants and fermented crickets are so flavorful that you know you are eating something special. "It is like exploding in your face. If it makes sense for the flavor, I would not mind at all serving them whole." Eating insects has environmental benefits as well. "I like that, apart from serving food, it is good for the environment. You should consider these things when you order your food, but again, from a restaurant perspective, we are always searching for delicious flavors. We are not a sort of model for a sustainable choice in restaurants. We established the food lab because I am interested in such topics, and once we started looking into this, insects came into the picture. That is one of the reasons for studying insects at the food lab: as an alternative to other protein sources for sustainable reasons."

DELICIOUSNESS

Yet nutritional and environmental issues are not the most important to Redzepi. "Unless the world is coming to an end and there is nothing else to eat, if it is not delicious, you don't get people to eat it. If it is delicious, it is a whole different story. We expect a meat crisis, in particular with beef, and some will adapt by eating chicken; but as a general acceptance of natural delicious foods, I think if it is not delicious, it becomes very complicated. You need to work hard and long for people to accept insects on their plate. You can persuade people such as hard-core NGO types to eat insects because it is good for planet Earth. However, the general mass of people out there, who are looking for quality of life, need a delicious mouthful. It is all about deliciousness, deliciousness, and deliciousness."

The chef is convinced that insects will innovate food culture. "It is the start of something new in cooking. I think it is going to be explored. We will always have it on the menu, because they are too good not to eat. Maybe there are some problems, but these are minor things. Testing, checking out, food safety, you need to have all that accounted for, but one can do that. It is a new beginning. A lot of people do not know what to do, so you invest time and money. In these days of technology, it will not be difficult to determine the food safety issues. We do not serve crickets to people with a seafood allergy, just to be safe. Yes, sure, it is a new world; you put yourself on the line when you explore new things. But that is exciting."

INNOVATION

Noma gets a fair amount of criticism for doing this. "A lot of people joke and make fun of us. We are in a very conservative business, and insects provide a whole new world of how meals are going to be served. People will say, 'There is a whole French history of what you eat, and suddenly it is ants? What is going on? You are just looking for publicity; there can be no way that this tastes good.' I felt the animosity toward people who eat insects, the Western self-righteousness. As if it cannot be good because people from 'lesser societies' eat these things. We found these three insect species—two ants and a cricket—that taste so good. We studied them; we found out that this could be something very positive for sustainability issues around the world. If we could replace 5 percent of our protein intake with insects, that would be a revolution, an ecological sensation. Okay, cool, wow. This we need to study. This we need to understand more. We see it as our job to make it very delicious."

KATJA GRUIJTERS, FOOD DESIGNER

"The Next Generation's Shrimp Cocktail"

In her office under elevated train tracks in Amsterdam, she serves chamomile tea with a sugared piece of licorice root to stir it. Within reach for snacking are dried cherries and blueberries. Food designer Katja Gruijters knows all about food and drink. And yet she wonders why a shrimp cocktail makes our mouth water but we think deep-fried grasshoppers are weird.

That shrimp cocktail is a fascination. "Who didn't grow up with that? We were bombarded with them in our youth. Even though shrimps resemble insects, we consider a grasshopper salad strange." Why is that? "I've eaten grasshoppers myself. It's very ordinary in Thailand." And to Katja Gruijters, actually, eating insects is indeed very ordinary. How can you get people to see eating insects as normal as eating shrimp is? In other words: How do you make them appealing to a large audience?

ENDEARING

"If I were to give you a grasshopper right now, you would not say, 'Oh, yummy.' A great deal will need to happen before you react that way. We have to work on the identity of the grasshopper. To me, it's a very likeable creature; just think of Flip the Grasshopper in *The Adventures of Maya the Bee*, or Jiminy Cricket from *Pinocchio*. You can make use of that likeability. It's not gray or creepy. It's quite endearing."

The world, or at least the *Western* world, has to be rid of its prejudices about insects. Gruijters warns, "We have to steer clear of hype or silliness. Not talk about the legs and things, but about why it's a good thing to eat insects."

SOCIAL ISSUES

The designer looks to the social arena to answer her question. "Isn't it bizarre that food rots in our refrigerators, that we no longer know how to cook, and that children don't know where their food comes from? That we have a surplus here and a shortage there? If we keep going on this way, while the world's population keeps growing and growing, we're going to have problems. We are so smart, but the extremes are just getting bigger. We haven't found a way to solve that problem. We should be thinking and acting in a different way when it comes to food and drink. I see it as my task to make consumers and businesses aware of their own behavior. I'm truly concerned about these problems, and as a designer, I can offer solutions."

Insects might be able to contribute to that solution—on account of their abundant protein. According to Gruijters, insect consumption is the most meaningful if it happens on a large scale, even internationally. "If you can't make it accessible to a large audience, it will remain a niche product. The important thing about insects is that they are a very rich, sustainable source of protein. That fact can help to cut back meat production, and that ties into the global hunger issue."

STREET FOOD

"In my work, I go looking for the source of an ingredient, and then for its essence. With insects, you end up in Asia, where grasshoppers, mealworms, and crickets are eaten as street food. How can you translate that given to the here and now? It can be done. You can offer it as a snack. At the farmers' market, or through a chain." Gruijters sees plenty of opportunities for fast-food concepts.

SPICE MIX

Even the sight of an intact animal can be enough reason for some people not to eat everything on their plates. It's like the head of a fish: cut it off, and there's a ready market for the fillet. "You can mix insects with other ingredients to make them tastier or more colorful." Gruijters suggests a spice mix in a transparent mill. One with a whole insect inside, because as far as she's concerned, it need not be concealed. There's no reason not to show it. "And you can market it globally. If you make chocolates with a grasshopper on top, you won't achieve your long-term goals. The spice mix actually serves to add protein and taste to a main meal such as a salad, or a delicious vegetarian quiche."

It's not only nutritious, but also nice and crunchy. "That's what's great about insects. If you deep-fry them and add herbs, you get a delicious 'bite' that you often miss in vegetarian food. I would use the spice mill with insects myself." She laughs. "Even though it's a little sadistic to grind up insects with your own hands. But it is good, because it gives us back the contact with our own food and drink just as it really is.

Model of herb mix and mill (Studio Katja Gruijters). (Lotte Stekelenburg)

Another use for insects is as a kind of meat substitute in our meals. The disadvantage of that is that it doesn't fit in with today's trend of wanting to know what our food looks like." In this form, you see the product but you don't recognize it as you eat it.

PACKAGING

And the packaging for the bugs themselves? What do you do with that? "Packaging is communication." And there's a lot to communicate when it comes to insects. "You're faced with the problem that no one is really familiar with them. So again, you have to tell the whole story." What kinds of bugs there are. Where they come from. "And don't show any creepy legs. Everyone should say, 'Hey, this is exciting. That looks tasty. I do want to try it.' And then you have to offer information on how to prepare the bugs."

The possibilities are endless and exciting, and also vital. If it's up to Gruijters, the next generation—and the next, and the one after that—really will grow up with insects on their plates.

Spiders in Cambodia

Tarantulas are found mainly in the tropics. Most of these spiders are 2 to 3 inches (5 to 8 cm) long or 7 inches (18 cm) including legs, though a few species can reach 1 foot (30 cm) in length. Tarantulas are covered with hairs that can irritate our eyes and mucous membranes. Although their bites can be very painful, most tarantulas are not dangerous for humans unless people happen to be allergic: only about 1 percent of tarantula species produce a venom harmful to humans.

The city of Skuon, about 45 miles (75 km) from Cambodia's capital, Phnom Penh, is the place to be for eating tarantulas. Legend has it that during the terrifying Khmer Rouge regime, people were so starved that they started eating spiders. Even after Pol Pot's regime was overthrown, the people of Skuon continued eating tarantulas because they are so delicious.

Skuon lies on a highway to the capital, and it is now known as the country's

Deep-fried spiders at a market in Skuon, Cambodia. (Mat Connolley)

center for this outlandish delicacy. People travel from the capital especially to buy tarantulas. As one tarantula seller explained, "Actually, people who aren't used to eating spiders don't really dare at first, but once they have tried them, they love them." She sells 100 to 200 spiders a day, at about six cents apiece. Earnings of six to twelve dollars a day don't seem like much, but in a country where the average daily wage is sixty cents, this is a small fortune. Not only the vendors earn on the tarantulas; the men who dig them up also profit. Tarantulas live in tunnels underground, like moles. Tarantula hunters poke a stick in the spider's tunnel and wait for the spider to attack; then they pull out the stick with the spider hanging on. A good hunter can catch a few hundred spiders per day.

Crunchy Tarantulas

Tarantulas
Sugar
Salt
Oil, for frying
A few garlic cloves, peeled and rushed

- Sprinkle the spiders with a mixture of sugar and salt.
- Heat the oil in a skillet, then add and stir-fry the garlic.
- When the oil is fragrant, add the spiders and fry until the legs are crisp and the abdomens are firm.

How does this dish taste? The legs are crunchy. The inside of the abdomen looks like white meat, with a taste somewhere between chicken and codfish.

Moths in Italy and Australia

When people say, "I have butterflies in my stomach," we never ask how those insects tasted. Yet butterflies and moths are eaten, albeit less often than caterpillars. In Carnia, in the north of Italy, until the mid-twentieth century, children were known to eat moths during the summer. More specifically, they ate the sweet contents of the crop of *Zygaena* moths. The crop, present in some insects and in birds, functions as a temporary reservoir within the insect's digestive system.

Zygaena are brightly colored insects, and that is normally a signal to predators that they are poisonous. This species indeed has in its body a natural type of cyanide probably acquired from plants. Early in the summer, however, the cyanide content is low. Because of their warning coloration, *Zygaena* sitting on flowers do not easily shy; the children in Carnia could easily catch them with their hands. By gently pulling the abdomen apart, they would expose the conspicuous crop. The deliciously sweet contents would be swallowed on the spot. The children didn't have to worry about cyanide poisoning—they would have had to eat 5,000 of these moths for it to be dangerous.

The moth *Zygaena transalpina*. (Maurizio G. Paoletti and Angelo L. Dreon)

The crop of the moth *Zygaena*. (Maurizio G. Paoletti and Angelo L. Dreon)

The Aboriginals in southeast Australia also eat a certain type of moth, the Bogong moth (*Agrotis infusa*). This moth can be found in large numbers in the Bogong High Plains, where it spends the dry summer season in rocky crevices. Since early times, Aboriginal people belonging to various tribes have chosen this area for special ceremonies because the Bogong moths are plentiful enough for a ceremonial feast. The moths are smoked out of the crevices with torches and collected on kangaroo skins. These nutritious insects contain about 50 percent fat. Roasting them over a fire burns off the scales, wings, and legs, and they can then be eaten as is or ground into a paste and shaped into cakes.

SIX DESSERTS

Chocolate Cupcakes. (Floris Scheplitz)

Chocolate Cupcakes

Equipment needed:
12-unit cupcake pan lined with paper cupcake liners

- 9 tablespoons (125 g) unsalted butter
- 1 cup (125 g) all-purpose flour
- 3 tablespoons unsweetened cocoa powder
- 1½ teaspoons baking powder
- ⅔ cup (125 g) brown sugar
- 2 large eggs
- 6 tablespoons milk
- 1 apple, peeled, cored, and finely diced
- ⅔ cup (25 g) buffalo worms

Preheat the oven to 350°F (180°C). Have all the ingredients at room temperature.

- Sift the flour, cocoa powder, and baking powder into a medium bowl.
- In a separate medium bowl, beat the butter with an electric mixer on medium speed until fluffy, adding the brown sugar gradually. Beat in the eggs one by one.
- Add the dry mixture in batches, beating well. Add the milk and beat for 1 minute more.
- Fold in the apple and buffalo worms with a spatula.
- Use a spoon to divide the batter among the cupcake liners, filling them two-thirds full.
- Bake the cupcakes for about 20 minutes, until a toothpick inserted into the center comes out clean.

YIELD: 12 CUPCAKES

Chocolate Cupcakes. (Floris Scheplitz)

Buglava. (Floris Scheplitz)

Buglava

Equipment needed: 8-inch (20 cm) square baking dish and pastry brush

8 ounces (250 g) frozen phyllo dough
Scant ½ cup (60 g) pecans
Scant ½ cup (60 g) blanched almonds
⅔ cup (25 g) mealworms
2½ ounces (75 g) almond paste
1 large egg, lightly beaten
2 teaspoons ground cinnamon
7 tablespoons (100 g) unsalted butter

SYRUP

6 tablespoons honey
Juice of ½ lime or lemon
¼ cup (50 ml) orange juice (bottled or fresh-squeezed)

Allow the phyllo dough to defrost in its package in the refrigerator. Preheat the oven to 300°F (150°C).

- Roast the pecans, almonds, and mealworms in a baking dish in the middle of the oven for 15 minutes. Remove from the oven and allow to cool.
- Chop the nuts and mealworms coarsely on a cutting board or in a food processor.
- In a medium bowl, gradually incorporate the egg into the almond paste, then mix in the cinnamon, chopped nuts, and mealworms.
- Increase the oven temperature to 350°F (180°C).
- Melt the butter in a saucepan and use a little bit to grease the baking dish.
- Cut the stack of phyllo leaves in half, keeping the dough slightly damp according to the package directions. Leaf by leaf, place one-third of the phyllo in the prepared baking dish and brush every leaf with melted butter. Distribute half of the mealworm mixture over the first layer of phyllo dough. Repeat this process, placing another third of the phyllo leaves on top, again brushing each leaf with butter, then distributing the rest of the mealworm mixture over it. Cover with the last third of the phyllo leaves, again brushing with butter.
- Sprinkle 4 tablespoons of water over the pastry and, while still in the dish, cut it into squares of equal size. Bake the buglava in the middle of the preheated oven for 35 minutes, until nice and brown.
- In the meantime, in a small saucepan over low heat, heat the honey with the lime juice, orange juice, and ¼ cup (50 ml) of water to make a syrup. When the buglava comes out of the oven, pour half of this syrup over it and allow it to soak in. After 30 minutes, pour the rest of the syrup over the buglava, then let it sit for 2 hours to allow the flavors to blend.

YIELD: SERVES 8

Tarte Tatin. (Floris Scheplitz)

Tarte Tatin

Equipment needed: 9-inch (23 cm) pie plate

CRUST

⅔ cup (150 g) butter
⅓ cup (75 g) superfine sugar
1 large egg, lightly beaten
Pinch of salt
2½ cups (300 g) all-purpose flour

SYRUP

⅔ cup (150 ml) very strong brewed coffee
2 tablespoons brown sugar
¼ teaspoon ground cloves
¼ teaspoon ground cinnamon

FILLING

8 grasshoppers, legs and wings removed
6 tart apples
3½ tablespoons (50 g) unsalted butter
½ cup (100 g) sugar
½ cup (20 g) mealworms, chopped

4½ ounces (125 g) bittersweet chocolate

1 quart (1 L) vanilla ice cream

Preheat the oven to 400°F (200°C).

- Prepare the crust: Using an electric mixer on medium speed, cream the butter with the sugar, egg, and salt. Attach the dough hooks to the mixer and knead the flour into the butter mixture until all the flour has been incorporated and the dough becomes sticky. Gather the dough together and continue to knead briefly by hand. Shape into a ball and allow to firm up in the refrigerator for 30 minutes.
- Prepare the syrup: In a pan, bring the syrup ingredients to a boil, lower the heat, and simmer for 5 minutes.
- Prepare the filling: Add the grasshoppers to the hot syrup, then allow it to cool, removing the grasshoppers after 10 minutes. Reserve the syrup.
- Peel and core the apples and cut them into eighths.
- Grease the pie plate with the butter and sprinkle with half of the sugar.
- Arrange the apple pieces in the pan, round side down, overlapping them as you go. Sprinkle with the rest of the sugar and the mealworms.
- Using a rolling pin, roll the dough out to a circle the size of the pie plate and place the dough over the pie plate so that it covers the apple filling.
- Bake the pie in the middle of the oven for 35 minutes, until golden.
- Melt the chocolate in a small bowl over, but not in, hot water. With a fork, pick up the grasshoppers one by one and coat them in chocolate, then place them on a plate in the refrigerator to harden.
- Allow the pie to cool in its pie plate. When cooled, place a platter upside down on top of it and turn them over together so that the pie is left resting on the platter.
- Cut the pie into wedges, serve with a scoop of ice cream, drizzle with a little reserved syrup, and garnish with the chocolate grasshoppers.

YIELD: SERVES 8

Chocolate Cake

Equipment needed: 9-inch (23 cm) cake pan and parchment paper

CAKE

5½ ounces (125 g) semisweet chocolate (70% cacao), chopped or broken up

9 tablespoons (125 g) unsalted butter, at room temperature

6 large eggs, separated, at room temperature

1¼ cups (150 g) powdered sugar, sifted

1 teaspoon pure vanilla extract, or the seeds scraped from a vanilla pod

½ cup (100 g) granulated sugar

1¼ cups (150 g) all-purpose flour

2 teaspoons baking powder

¾ cup (30 g) buffalo worms

FILLING

1 cup (250 g) apricot jam

3½ tablespoons coffee liqueur or chocolate liqueur

Preheat the oven to 355°F (180°C). Line the bottom of the cake pan with parchment and grease the sides of the pan.

- Prepare the cake: Melt the chocolate in a small bowl over, but not in, hot water.
- In a large bowl, beat the butter with an electric mixer on medium speed for 2 minutes. Add the powdered sugar and continue to beat until fluffy. Beat in the egg yolks one by one, adding the next one when the mixture has again become light and fluffy. Add the vanilla and the melted chocolate and beat for 1 more minute, until smooth.
- In a separate, large clean bowl, beat the egg whites on high speed until soft peaks form, then gradually beat in the granulated sugar and continue to beat. When stiff peaks form, fold one-third of the egg whites into the batter. Then add and fold in the rest of the egg whites.
- Sift the flour and the baking powder together over the batter and sprinkle the buffalo worms on top. Fold everything together, then spread the batter evenly in the prepared cake pan, smoothing the top.
- Bake the cake in the middle of the oven for 45 minutes, until a toothpick inserted into the center comes out clean.
- Turn the cake out of the pan onto a wire rack. When cool, slice the cake horizontally into two layers.
- Fill the cake: Heat the jam in a small pan. Sprinkle the two cake layers with the liqueur. Spread half of the warm apricot jam on one layer, and place the other layer over it. Coat the sides of the cake with the rest of the jam.
- Prepare the glaze: Bring the water, granulated sugar, and honey to a boil in a small saucepan over medium heat, then simmer for 5 minutes. Remove the pan from the heat, then add the chocolate pieces and stir well. Allow to cool slightly.

Chocolate Cake. (Floris Scheplitz)

GLAZE

⅔ cup (150 ml) water
¾ cup (150 g)
    granulated sugar
6 tablespoons honey
8 ounces (225 g)
    semisweet chocolate,
    chopped or broken
    up

DECORATION

2 ounces (55 g) almond
    paste, colored yellow
    with food coloring
1 large egg white
2 ounces (55 g)
    semisweet chocolate
A few blanched
    almonds

- Set a rack over a baking pan to catch the glaze drippings. Place the cake on the rack and pour the glaze over the cake, smoothing it with a large, flat knife. Make sure the top and sides are all well covered with glaze.
- Decorate the cake: Shape the yellow-tinted almond paste into little bee bodies with slightly tapered ends, and with small balls for the heads. Glue the pieces together with a bit of egg white. Melt the chocolate in a small bowl over hot water. Make a little decorating bag by folding a small piece of parchment paper, and fill it with melted chocolate. Draw eyes, mouth, and stripes onto the bees with the melted chocolate. When the chocolate has hardened, stick halved blanched almonds into the bees for wings.

YIELD: SERVES 8

Tips

- Use your imagination to decorate the cake. You might use grasshoppers or nuts, for example, or curls of chocolate.
- Instead of liqueur, you can sprinkle the cake with rum or a fruit syrup.
- A dollop of not-too-sweet whipped cream on the side is a nice complement to this cake.

Buffalo Snaps

Equipment needed: baking sheet and parchment paper

8 tablespoons (120 g) unsalted butter, at room temperature

⅓ cup (65 g) superfine sugar

1 teaspoon pure vanilla extract

Pinch of salt

Scant ⅔ cup (75 g) all-purpose flour, sifted

½ cup (50 g) cake flour, sifted

⅓ cup (15 g) buffalo worms

3 tablespoons (40 g) granulated sugar

- With an electric mixer on medium speed, cream the butter with the superfine sugar, vanilla, and salt. Attach the dough hooks to the mixer and add the flour and buffalo worms, continuing to mix on low speed until all the flour has been incorporated and the dough becomes sticky. Use your hands to gather up the dough and knead it briefly.
- Shape the dough into a ball and allow it to firm up in the refrigerator for 30 minutes.
- Divide the dough in two, and shape each half into a roll 6 inches (15 cm) long.
- Coat both rolls well by rolling in the granulated sugar, taking care that they don't get any longer. Set them on a dish and chill in the freezer for 30 minutes.
- Preheat the oven to 350°F (175°C).
- Cut the rolls into slices about ⅓ inch (1 cm) thick. Line the baking sheet with parchment paper, then arrange the cookies on it 1½ inches (4 cm) apart; bake for 20 minutes, until golden.
- Allow the cookies to cool on the baking sheet. Store in a cookie tin.

YIELD: 30 COOKIES

Buffalo Cinnamon Cookies

Equipment needed: baking sheet, parchment paper, and pastry brush

- **8 tablespoons (120 g) butter, at room temperature**
- **Scant ½ cup (85 g) superfine sugar**
- **1 large egg, lightly beaten**
- **1 teaspoon baking powder**
- **Pinch of salt**
- **1²/₃ cups (170 g) cake flour, sifted, plus more for rolling**
- **2 teaspoons ground cinnamon**
- **¹/₃ cup (15 g) buffalo worms**
- **¼ cup (25 g) sliced almonds**
- **2 teaspoons coarse sugar**

- With an electric mixer on medium speed, cream the butter with the superfine sugar, half of the egg, the baking powder, and the salt. Attach the dough hooks to the mixer and add the flour, cinnamon, and buffalo worms, continuing to mix until all the flour has been incorporated and the dough becomes sticky. Use your hands to gather up the dough and knead it briefly.
- Shape the dough into a ball and allow it to firm up in the refrigerator for 30 minutes.
- Preheat the oven to 350°F (175°C).
- Briefly knead the dough to soften it. Using a rolling pin and a little flour to prevent sticking, roll out to a rectangle of about 8 × 12 inches (20 × 30 cm). Line the baking sheet with parchment paper, then place the dough on it. Brush with the rest of the egg, then sprinkle with the sliced almonds and the coarse sugar. Bake for about 20 minutes, until lightly browned. Remove from the oven and cut immediately into bars about 1 × 3 inches (3 × 8 cm). Allow to cool, then store in a cookie tin.

YIELD: 28 BARS

Students of the Hotel and Tourism School (Rijn IJssel Vakschool), Wageningen—City of Insects festival, 2006. (Henk van Gurp)

4

On the Future and
Sustainability

Wageningen—City of Insects, 2006. (Frits Weener)

Mopane Caterpillars in Southern Africa

About 20 percent of all edible insects are caterpillars. Especially popular in southern and central Africa, one caterpillar species stands out among all the others. In southern Africa (Botswana, Namibia, Zimbabwe, and South Africa), in a 7,700-square-mile (20,000 sq. km) area—about the size of New Jersey or Wales—9.5 billion saturniid moth caterpillars are collected each year: more caterpillars than there are people on Earth! Their market value is approximately $8.5 million. The caterpillar feeds mainly on the mopane tree, which is why it is known as the mopane caterpillar.

The caterpillars are harvested when they are full grown and almost ready to pupate. They are hand picked from the trees, mostly by women and children who must sometimes travel hundreds of miles to areas where the caterpillars are found. After picking, the insects are pinched and squeezed to expel their gut contents, and then boiled in salted water and sun-dried. Prepared this way, the caterpillars can be stored

Market in the Democratic Republic of the Congo. (Giulio Napolitano FAO)

without refrigeration for months, just like other kinds of dried fish or meat. The dried insects can be soaked in water to plump them before cooking, but local people like to eat them dried as well, as a crispy snack.

Mopane worms are an important financial resource for the rural population, representing up to a quarter of the yearly income and allowing people to buy clothes, school supplies, and household goods. The caterpillars consist of more than 60 percent protein, and contain valuable essential fatty acids as well as such minerals as calcium, zinc, and iron. Their market value often exceeds that of beef.

Mopane caterpillar and moth. (Arne Larsen)

Mopane caterpillars at a market in southern Africa. (Kenichi Nonaka)

Mopane caterpillars at a market in southern Africa. (Kenichi Nonaka)

Mopane Caterpillar Stew

¾ cup (100 g) dried mopane caterpillars
1 onion, finely chopped
2 green bell peppers, seeded and diced
6 medium-size tomatoes, chopped
1 to 2 teaspoons curry powder

- Rinse the caterpillars and boil them in salted water in a medium saucepan for 30 minutes.
- Drain the caterpillars and add the remaining ingredients to the pan, as well as about 2 cups (500 ml) of water.
- Simmer gently, partially covered, for about 1 hour.
- Serve with millet or maize porridge.

Silk Moth Pupae in China

The story goes that silk was discovered by the Chinese empress Xi Ling-Shi (Leizu) 5,000 years ago. She found a silk moth cocoon on the mulberry tree in her garden and was able to pull a thread out of the

Dish prepared with silk moth pupae in Hangzhou, China. (Marcel Dicke)

cocoon and wind it around her finger, and then used this thread to weave cloth. For centuries, silk clothing was worn exclusively at the imperial court, and the Chinese were able to keep the exclusive secret of silk until about the year 600 c.e. Even the Roman historian Pliny the Elder got it wrong: in the year 78 c.e., he wrote in *Naturalis Historia* that silk grew as a downy layer on the leaves of an exotic Chinese tree, and that one could wash the silk off of the leaves with water. In China, revealing the secret of silk was punishable by death. Legend has it that monks finally smuggled some moth eggs and mulberry cuttings into Europe in a hollow stick, toward the end of the sixth century.

Silk moth cocoons are about 1 inch (2 to 3 cm) long, and are made up of a single thread that can reach up to 3,000 feet (900 m) in length. Two to three thousand cocoons are needed to make about 2 pounds (1 kg) of silk. When they are ready to pupate, silk moth larvae take about three days to spin the cocoon around themselves. With the yearly worldwide silk production, you could span 180,000 silk threads from here to the moon! Silk moths have been cultivated for thousands of years. The cultivated silk moth has been bred in such a way that the adult moth is no longer able to fly. If it were to be allowed to emerge normally from its pupa, the moth would make a hole in the cocoon, so it would

Silk moth pupae presented at an insect buffet in Thailand. (Arnold van Huis)

no longer be possible to unwind the silk thread. To prevent this, the cocoons are immersed in hot water before the moths can emerge. The silk can then be easily uncoiled.

But what to do with the leftover pupae? Eat them! Their protein content is about 60 percent, and they contain high amounts of unsaturated fatty acids, such as linoleic and linolenic acids. In China and Vietnam, people roast the cocoons on the barbecue; Koreans boil their cocoons. Boiled and canned silk moth pupae can be found at many markets in Asia, sometimes prepared in chili sauce. These canned pupae can also be barbecued or fried.

How do you prepare silk moth pupae? Just fry them in some hot oil. No extra salt or spices needed; they are tasty enough as is.

Silk moth pupae. (Arnold van Huis)

Food for Astronauts

How can you feed yourself on another planet, where nothing grows, or while traveling through space to get there? There is limited room inside a spacecraft, especially if you also need to produce food there. Some crops can be grown in small spaces, but producing meat, fish, eggs, and dairy products is more problematic. Japanese, Chinese, and Canadian scientists suggest using edible insects. Insects can convert plant material into high-quality food. This is possible in a small space, with robots handling the rearing process. Waste products, such as insect excretions as well as dead eggs, larvae, or pupae, can serve as nutrients for the plants. In this regard, insects are completely recyclable.

This team of scientists devised a system whereby 100 people can be provided with insect protein in space, as long as there is enough food for the insects. While a large number of the insects grown could be used for human food, the rest could be used to keep the colony going. As a model insect, the scientists chose the bread beetle, or drugstore beetle (*Stegobium paniceum*), of which the larvae are edible. They calculated that a 13 × 15-foot (4 × 4.5 m) room that is 8 feet (2.5 m) high would be sufficient to provide 100 people with their respective daily protein requirement.

Bread (drugstore) beetle. (Siga)

Silk moth larvae appear to be another good possibility. The Chinese researcher Yang Yunan, from Beijing University's Aerospace Department, believes that these insects will become the standard meal for Chinese astronauts because they are so rich in protein and easy to grow on leafy vegetables, using only small amounts of waste water.

Masamichi Yamashita, a Japanese scientist, agrees. At the Thirty-sixth Scientific Assembly on Space Research, held in Beijing in 2006, he presented a recipe for Savory Silkworm Cookies—not to be confused with space cakes—to be used during space travel.

Savory Silkworm Cookies

1½ teaspoon finely ground silk moths
1¼ cups (200 g) finely ground rice
1½ teaspoon finely ground soybeans
1¼ cups (300 ml) unsweetened soy milk
Soy sauce
Salt

- In a large bowl, stir the moths, rice, soybeans, and soy milk and season with soy sauce and salt to taste.
- Shape the dough into cookies on a microwave-safe baking sheet.
- Bake in a microwave oven for 15 minutes.

JAN FABRE, ARTIST

"I've Always Put Everything in My Mouth"

He calls himself a "consilience artist," and is known as a visual artist, a dramatist, and an author. In every discipline, he has started tongues wagging—or even caused great commotion, such as the time he wrapped the pillars at the entrance to the auditorium of the Belgian University of Ghent in slices of ham. He has created various installations incorporating insects; he covered the ceiling of a room in the Royal Palace of Belgium with millions of beetle elytra (hardened forewings). Where did Jan Fabre get his fascination for insects?

Apparently, that insect fascination runs in the family. Belgian artist Jan Fabre is a distant relative of famous French entomologist Jean-Henri Fabre. Yet, whereas the scientist was interested in only absolute truths, the artist dedicates his life to beauty in the broadest sense of the word. And in that pursuit, he often makes use of insects.

FRANKENSTEIN

"As a youngster, a seven-year-old little kid, I was fascinated by things that crept around, such as spiders." He laughs at the memory. "I studied them without a scientific method: grabbed the spider, pulled out a leg, and another leg, and another, to see what a creature like that would then do. I was a kind of Dr. Frankenstein, experimenting with little animals. I wasn't afraid of insects. I remember that I always teased my sisters with the spiders I had caught."

SCENT

"Later, I started a project in my parents' garden, about scent. I built a tent in the shape of two noses, in which I sat at a table with my microscope. I was planning to create new life from odors. I started with the scent, and brought body parts together, putting the wings of a fly on a worm, for example. I had developed this idea about how things should stink." Fabre is still doing this today. "Art is not only about form, but also about scent. Scent provides content." In other words: "Good art should stink."

The step from scent to taste is a small one. "Speaking of taste: I literally eat things up. I also teach my students to taste all of their materials. When people get older, they don't do that anymore, in contrast to children. That's a shame. I've always put everything in my mouth. How does copper taste? How does that color taste? And that brush? I am very physically engaged in making things. Even those spiders when I was seven. I've eaten everything. Even worms." Still, the Belgian artist is not wholeheartedly enthusiastic about insects on his plate. "Naturally, I know that they eat those beautiful little bugs in many cultures. So I've tried that myself, and even heated them up, but I'm not a big fan. I know that they contain a lot of protein and other nutrients, but still—no. I'd rather have a steak and fries." The innovative artist is averse to novelty when it comes to food. "There are restaurants that do—what do you call it?—molecular cooking. It's really expensive, but I don't think much of it."

Jan Fabre, homage to the famous entomologist Jean-Henri Fabre. Jewel beetle forewings on plaster, 2004, 68 × 47 × 33 cm. Installation view, *Messengers of the Death*, Salzburg (MAM Collection, Salzburg). (Studio Ghezzi © Angelos)

JEWEL BEETLES

"In 1992 or 1993, I made my first sculptures with real jewel beetles or jewel scarab beetles." Later, the artist switched to using only the insects' elytra. "There have been a few incidents. In the second half of the nineties, at the Venice Biennale—the queens of both the Netherlands and Belgium were present—larvae were dropping out of my sculpture. It turned out that a fly had laid eggs in it. The queen said, 'Something's alive. Is that normal?' Yes, because good art should die before the eyes of the viewers. All great art is preparation for death. Still, since then I did make the move from whole insects to only the forewings, for practical and hygienic reasons." For the world-famous ceiling art in the Mirror Room of the Royal Palace of Belgium in Brussels, for example. "If I had chosen to use whole beetles for that, surely a moment would have come when the queen or another distinguished person would find a larva in her champagne glass."

A MILLION AND A HALF

Whether you like it or not, that ceiling is an impressive work. "I used a million and a half jewel beetle forewings. I had arranged for them to be collected for two whole years at restaurants in Indonesia, Malaysia, and the Congo, where they are actually barbecued on skewers. The beetles resulted in a mosaic of light on the ceiling. In the drawings of that mosaic, I put in all kinds of references to our colonial history. I expressed criticism of that history, by using skulls, severed giraffe legs,

Jan Fabre, jewel beetle forewings on the ceiling of the Mirror Room, Royal Palace, Brussels, 2002. (Dirk Pauwels © Angelos)

disfigured animal species, and elephant tusks. As colonists, we actually plundered half of the Congo; we tortured people for diamonds and ivory. But internationally, the criticism was overlooked. It was only about how sublime that artwork was. How beautiful, how the colors changed. And of course it is breathtaking and overpowering when you see it in the light, especially in a palace room like that." It led Fabre to use jewel beetle forewings to make another work and again express his criticism of Belgium's colonial past.

SYMBOL

As severely as he speaks of that subject, he talks lovingly of insects. "I notice that the better collectors, museum directors, and curators see the beauty of it. They can place it in the context of Flemish and Italian art history. Yet the same eminent international art collectors say, 'My wife could not live with those insects; she thought it was disgusting.' That's why my work isn't hanging in their living room. We live in a society in which people take everything they don't know, and everything that doesn't make quick money, and sweep it under the carpet. I see the beetles as a bridge between life and death. And death, then, is something positive, an energy field."

It says something about our society that we are afraid of these fantastically beautiful little beings. And that we think we can do without them, which of course is not true. I always tell people that insects live in every healthy house. A house is not healthy without insects."

Shellac from India

First, we eat with our eyes. When shopping, we select fruits and other foods that look unblemished and glossy. Just take a look in a store at how many foods sparkle at us. Shiny things are attractive; the implication is that they are pure and wholesome. The food and candy industry makes clever use of this. Fruits are usually treated to make them shine, and candies and chocolates are often coated with a special finish as well.

Shellac. (Ivory)

Shellac (food additive number E904) is an important glazing agent used in the food industry. It is a natural product that comes from the lac scale insect (*Laccifer lacca*) from India. This insect feeds on tree sap and is related to aphids and the cochineal scale insect. Shellac is a substance secreted by the lac scales, and these insects use it to make a protective shell. Branches are often completely covered with shellac. It can be scraped off, then dried and purified.

Shellac is used not only as a glazing agent on fruits and candy, but also as a coating on timed-release medications because it is resistant to gastric acids. In the past, shellac was also one of the components of 78-rpm phonograph records, before the advent of vinyl.

Jumping Plant Lice in South Africa and Australia

Both the Bible and the Qur'an refer to manna: the heavenly bread that would rain down from the sky, and that could be gathered up from the ground but not kept for very long. Manna was perceived as a gift from God. Biologically speaking, manna is the sweet substance also known as honeydew, which comes from plants and is exuded as a sticky residue by aphids, and also by jumping plant lice.

There are examples of this honeydew manna in South Africa and in Australia. It is called mopane bread in South Africa, because the honeydew is produced by jumping plant lice, also called psyllids, living on mopane trees. The sweet secretion dries up into small, hard, conical, and transparent structures also known as *lerp*. This term actually describes not only the manna, but the entire combination of the insect, the scales it sheds, and its secretions. Locals gather up the *lerp*, grind it into flour, and store it for when other food is scarce. *Lerp* can be collected only in the dry season, because rain washes it off the leaves. It can make a good meal when mixed with milk into a kind of porridge. Mopane bread has 250 calories per 3.5 ounces (100 g) and is high in carbohydrates, potassium, and phosphorus, but has little protein. Two other *lerp* consumers—the mopane caterpillar and the elephant— compete with humans. Both of these animal species are very fond of mopane leaves.

The Australian Aboriginal people use *lerp* from an Australian species of jumping plant lice, as a sugar source. This insect lives in eucalyptus trees. The Aboriginals gather the conical *lerp* structures, which are 2 millimeters high and 4 millimeters in diameter, from the affected leaves. Sometimes they even chop down an entire tree so as to facilitate the *lerp* harvest.

Dirck Metius (seventeenth century, Alkmaar), *Willem van Loon and His Family as Israelites Gathering Manna* (Exodus 16:17–31). Royal Netherlands Academy of Arts and Sciences (Koninklijke Nederlandse Akademie van Wetenschappen), Amsterdam.

Insects: A Sustainable Alternative to Meat

The Food and Agriculture Organization of the United Nations (FAO) estimates that livestock is responsible for 18 percent of all greenhouse gas

Insect trap near Luang Prabang, Laos. Insects, attracted by the lamp at night, fly into a sheet of aluminum and fall into a tub of water. This is how the edible insects are gathered. (Arnold van Huis)

emissions, and is, as such, an important contributor to global warming. Greenhouse gas emissions include methane (CH_4) and nitrous oxide (N_2O). Simply by burping and passing gas, cattle release more than one-third of all methane emissions worldwide. Methane contributes twenty-three times more to global warming than does carbon dioxide (CO_2), the most important greenhouse gas emitted by cars. Livestock generates close to two-thirds of all nitrous oxide released; this gas is 289 times more damaging than CO_2. Some insects, such as termites, also produce methane, releasing 4 percent of all emissions of this gas worldwide. By contrast, the edible insects mentioned in this cookbook, such as mealworms and migratory locusts, produce far less greenhouse gas per kilogram of product than do cows or pigs.

Livestock also produces more than two-thirds of the world's ammonia emissions, which are one of the main causes of acid rain. Per kilogram of body weight produced, pigs produce fifty times more ammonia than do locusts.

Insects are cold-blooded; they do not need to metabolize food to maintain a constant body temperature. That is why insects are so efficient at converting feed to an edible product. A mere 4.6 pounds (2.1 kg) of feed is sufficient to produce 2.2 pounds (1 kg) of edible crickets, whereas 55 pounds (25 kg) of feed is needed to produce 2.2 pounds of beef.

Marieke Calis in the mealworm-rearing room. (Lotte Stekelenburg)

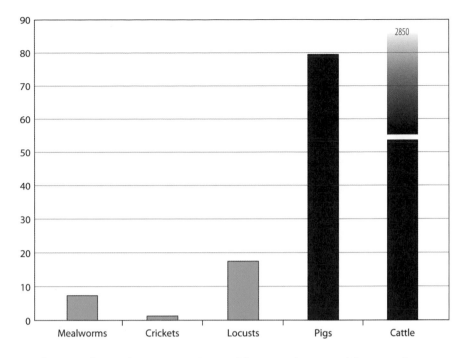

Production of greenhouse gases (gram CO_2 equivalents per kilogram of growth) by edible insects, compared with pigs and cattle.

	Crickets	Poultry	Pork	Beef
Feed conversion ratio (kg feed/kg live weight)	1.7	2.5	5	10
Edible portion (%)	80	55	55	40
Feed (kg/kg edible weight)	2.1	4.5	9.1	25

Production efficiency of edible crickets, compared with poultry, pork, and beef. (Adapted from Arnold van Huis, "Potential of Insects and Food and Feed in Assuring Food Security," *Annual Review of Entomology* 58 [2013]: 563–83)

As well as looking at what an animal itself produces in terms of greenhouse gases, it is also interesting to consider all environmental impacts associated with producing conventional meat or insects. For example, we can take into account the environmental impact of producing the fodder for the animal (sowing, irrigation, fertilizing,

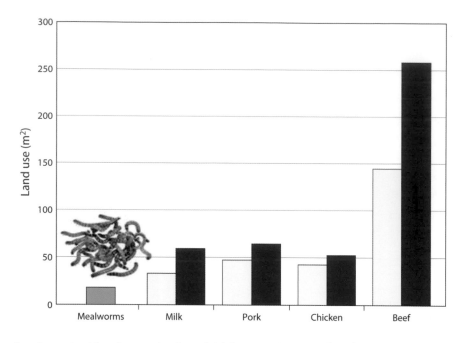

Land required for the production of 1 kilogram (2.2 pounds) of edible protein. ([*green bar*] Data from Dennis G. A. B. Oonincx and Imke J. M. de Boer, "Environmental Impact of the Production of Mealworms as a Protein Source for Humans: A Life Cycle Assessment," *PLoS One 7*, no. 12 [2012]: e51145; [*yellow and blue bars*] minimum literature data)

transporting the feed to the animal, the energy used at the farm, and the like). This has recently been analyzed for mealworms and compared with milk, pork, chicken, and beef production with respect to three parameters: energy use, greenhouse gas production (global warming potential), and land use.

The energy used to produce 2.2 pounds (1 kg) of protein from mealworms is lower than that for beef, comparable to pork, and slightly higher than for chicken and for milk. The greenhouse gas emissions resulting from mealworm production are much lower than for the more commonly farmed animals. For the production of milk, pork, and chicken, more land is required than for the production of mealworms, but to produce beef, ten times as much land is needed. The amount of land needed for production is an important factor to consider, as two-thirds of the agricultural land on our planet is already being used for livestock.

HERMAN WIJFFELS, POLITICIAN

"A New Episode in the History of Our Civilization"

Although he and his wife very occasionally allow themselves a piece of meat, Dutch economist Herman Wijffels has been a confirmed vegetarian for twenty years. "I feel good about that." In his view, insects fall into a category between conventional animal proteins and pure vegetarian ones. Yes, he is open to eating them. "Most likely, it can even increase the diversity in our diet." But it turns out there are more reasons for having insects on the dining table.

Eating insects is not only exciting or delicious—it's also necessary. Maybe even inevitable. To understand why, Herman Wijffels goes back to the basis of our economic system. In any case, the way we produce food right now can't go on for long. "We are running up against physical limitations. It hasn't quite gotten through to us yet, but for the first time in the history of humanity, we are in the predicament of exhausting our natural resources. We call it overshoot: we are overtaxing our fresh water sources, our agricultural land, and the ocean's fish supplies." As a former banker, he compares the food industry to the financial market. "For part of the year, we're living on the principal of our natural capital, instead of on the interest. If you want to live sustainably, you can't take more out of the natural resources than what is being added to them. But that is happening, and on a large scale."

CYCLE

"That's why we are now working on the development of new economic processes. From the old *linearly* organized economic processes toward *circularly* organized processes. The most prominent characteristic of circular is much more efficient management of such natural resources as fossil fuels and minerals." In practice, for example, you strive to reuse instead of to destroy. "At the end of a product's life span, you don't throw it away; you use the raw materials again for subsequent production. Cycle upon cycle upon cycle."

"My whole way of thinking at this point is infused with the idea that every investment we make should reduce our environmental footprint. At the global level, our footprint is one and a half times the structural carrying capacity of our planet. The vision we work with is to replace processes that create a bigger footprint with processes that reduce it. That's how we evaluate every innovation."

CONVERSION FACTOR

What does that have to do with insects? A great deal. Because they can be reared incredibly efficiently, whether abroad or just around the corner.

Wijffels: "I read in the newspaper that a third of the farmland in the world is showing signs of a serious degree of exhaustion. Our methods of working the land are systematically harming the soil. We should be harvesting responsibly, and making use of the possibilities that biological circumstances offer. Insects require little space. Furthermore, they have a high conversion factor. That is, you don't have to give them much to eat in order to yield a high nutritional value. For just 2 pounds [1 kg] of beef, you also need about 10,600 gallons [40,000 L] of water. I don't think we can keep this up. The capacity for that just doesn't exist."

COLD BLOODED

For insect production, though, there does appear to be sufficient capacity at the global level. The conditions for insects to grow are most favorable in the tropics. Insects are cold-blooded: they take on the temperature of their environment. And they develop better at the high temperatures of the tropics than in temperate zones, especially during our cold winter. "Farming insects in the tropics and transporting them to Europe fits into the circular way of thinking."

Then again, Wijffels also sees possibilities for farmers in the Netherlands to rear insects, just as there are now farmers who have begun producing algae. "As long as it is economically feasible for them. A big advantage to insect production is that you break the cycle of importing nutrients from overseas that creates a surplus here and a shortage there. You can make use of local organic waste materials for the production of insects: you close the loop. It fits perfectly into a circular economy." Because, by using residual streams as food, you create a closed circuit.

DETOUR

"Using vegetable materials to produce animal proteins is a detour. Strictly speaking, everything we need in order to live is found in plants. I expect that we will be avoiding that detour more and more. According to the WHO [World Health Organization], the effects of overconsumption of animal proteins on health are not very good. In a balanced diet, there is less room for animal protein. So the market for animal proteins will be shrinking. You have to take that into account when calculating the feasibility of insect production."

But will animal protein consumption increase in such countries as China and India? Does the growing Chinese population mean, by definition, that we should be seeing a shortage of animal proteins? Not necessarily. "For the sake of convenience, people assume that the same developments will occur there as here. That may be a possibility in China, but not in India. The Indian diet is much more vegetarian. If you go to an Indian restaurant, the emphasis is often on vegetables. Meat plays a much less dominant role than in Western cuisine. So I hesitate to directly translate the development to 'what has happened here will also happen there.' But that the growing prosperity in China and India will mean an increased demand for animal proteins in the coming decades seems quite obvious. The question is how that will play out in terms of prices. If the Chinese triple their consumption of sweet-and-sour pork, they certainly won't make it." There's too little livestock, as well as a shortage of livestock feed. "The way the future looks now, the total world trade volume of grain and soy would go to China. Naturally, that can't happen."

INFANCY

"I don't exclude the possibility that new market niches can be developed in which animal protein based on insects—with their favorable conversion factor—is a better option than protein with an unfavorable conversion factor. More research on that is needed. Actually, it's crazy that so much research is done in the food industry, yet exploiting insects as an animal protein source is still in its infancy."

The crude insect product can be processed. But before the food industry truly sees the point of these "new" products, Wijffels suspects, enormous volumes of insects will have to be available. But then again, if the conversion factor is only 4.6 pounds (2.1 kg) of feed to 2.2 pounds (1 kg) of edible body weight gain, and insects can be grown on organic residual streams, "then it could be done," the economist declares.

FUTURE

Does he really think it will happen? "Why not? I think that the average Dutchman would rather have a pork chop on his plate than an insect. But in other parts of the world, people enjoy eating insects." And fortunately, the Dutch sometimes allow themselves to be seduced into a little culinary development. "Just look at how international food culture—and certainly not just Chinese food—has found its way into Dutch culture. We stopped eating only meat and mashed potatoes a long time ago."

Attendees tasting tidbits at an insect-cooking competition in Laos. (Thomas Calame)

PAUL VANTOMME, FAO OFFICIAL

Insect Consumption: A Global Perspective

The FAO predicts that by the year 2050, 70 percent more food will need to be produced, to feed the world's growing population. Paul Vantomme, senior forestry officer of the FAO, explains why insects are of global importance as a source of protein.

Our current food production methods will have to become more efficient. Furthermore, "new" foods, such as algae, seaweeds, or insects, will need to be developed. Our production, processing, and distribution systems will also have to become more sustainable. We will have to take into account the changing growing conditions related to climate change and water scarcity. These are reasons why the FAO is studying how insects can play a part in food security.

Girl tasting tidbits at an insect-cooking competition in Laos. (Thomas Calame)

Insects are already a delicacy for some 2 billion people. The majority of insects that are now eaten worldwide are harvested in the wild. For the most part, this harvesting takes place in the tropics—especially in forests, to reduce the risk of exposure to pesticide residues. "Farming" insects occurs on a small scale, predominantly in developed countries, and is usually for the purpose of feeding pets or fish. In such countries as Thailand, Laos, and China, insects are also grown for human consumption. In terms of technique, catching or farming insects is relatively simple. In developing countries, women and children are usually the ones to gather insects, as a supplement to their own meals. Women, as well, are primarily the ones who sell dried or prepared insects at the local market. For many families, then, growing and selling insects is an essential source of income. Insect farming has the added advantage that it is not dependent on owning land and does not require expensive equipment. Grasshoppers, for example, can be grown in an old oil drum, even on the balcony of an apartment in the city, or under a bridge. What's more, insects can be fed with leaves or other organic material that can usually be collected for free, even in the cities. In Thailand, for example, about 20,000 families are farming crickets this way, and then drying and selling them as feed for chickens and fish, and also for human consumption.[1] Thanks to these activities, the monthly income can reach as much as $120, which is a lot of money in a country where the average monthly income is $80 per person. Insect farming can thus help combat poverty, in cities as well as in the countryside.

Before insects become a large-scale, nourishing food product for humans and animals worldwide, much is yet to be done. In the Western world, we need to overcome culturally determined aversion. Additional research is also needed to make insect farming more cost-effective. Most important, insects must be turned into an appealing product. Food based on insects will, after all, have to compete with traditional and inexpensive meat products. And last but not least, insect-based dishes have to taste delicious; hence this cookbook as a source of inspiration for adventurous and innovative cooks.[2]

1. Yupa Hanboonsong, Tasanee Jamjanya, and Patrick B. Durst, *Six-Legged Livestock: Edible Insect Farming, Collection and Marketing in Thailand*, RAP Publication 2013/03 (Bangkok: FAO, Regional Office for Asia and the Pacific, 2013), http://www.fao.org/docrep/017/i3246e/i3246e00.htm.

2. For more on edible insects, see Food and Agriculture Organization of the United Nations, "Edible Forest Insects," http://www.fao.org/forestry/edibleinsects/en/; and Arnold van Huis et al., *Edible Insects: Future Prospects for Food and Feed Security* (Rome: FAO, 2013), http://www.fao.org/docrep/018/i3253e/i3253e00.htm.

Insect Consumption: The Future

Since 1997, we at Wageningen University have been drawing attention to the fantastic new possibilities that eating insects can bring. At first, the reactions were characterized—understandably—by disbelief. As the years have gone by, however, it has become clear that the future looks bright for insect consumption; people have been increasingly positive and interested. One example is an article that appeared in the Dutch Railways magazine *Tussen de rails* (*Between the Rails*) around the turn of this century, about trends we would be seeing in the new century—one new phenomenon was described per year. For 2016, the prediction was the opening of the first insect fast-food restaurant, MacCricket.

Interest in eating insects has skyrocketed since 2000. In 2005, former Dutch minister of agriculture Cees Veerman spoke of the need to find alternative animal protein sources for the incessantly growing world population. In 2006, the entomologists at Wageningen University transformed the town of Wageningen into the "City of Insects," an event including an attempt to set a world record for insect consumption. Simultaneously, a total of 1,747 people crowding the city's main square ate mealworms prepared and distributed by professional cooks. Several Dutch companies have started growing insects for human consumption, and these companies also founded the trade organization Venik (Verenigde Nederlandse Insectenkwekers [Dutch Insect Farmers Association]). As a result, insects are now commercially available in the Netherlands through the wholesaler Sligro, and can also be bought from online retailers (see page 180). In 2008, the Horecava (annual food exhibition) presented insects as *the* food of the future—9,500 visitors tasted them. Former Dutch minister of agriculture Gerda Verburg served up an insect meal to her European Union colleagues in 2010. The consumer program *Radar* questioned visitors to its Web site in May 2010, regarding their views on eating insects; a surprising 30 percent said they would be willing to try. A 2011 contest voted in the favorite new word in the Frisian language: *hipperhapke*, a translation of the Dutch term for "insect snack." The same year, Marcel Dicke and Arnold van Huis, two of the authors of this book, were invited to write an article for the *Wall Street Journal* about insect consumption in the Western world.[3]

3. Marcel Dicke and Arnold van Huis, "The Six-Legged Meat of the Future," *Wall Street Journal*, February 19, 2011, http://online.wsj.com/article/SB10001424052748703 2932045761060723400200728.html.

In 2013, the book *Edible Insects: Future Prospects for Food and Feed Security*, coauthored by Arnold van Huis, was downloaded 2.3 million times in the first twenty-four hours after its launching—a demonstration of worldwide interest in this topic.

An article in the December 24, 2010, issue of *New Scientist* presented a "retrospective" as told from the year 2050.[4] This article recounted that, in 2030, insects had become a breakthrough solution to the meat crises that had afflicted the world for years. The author of the article pointed to the many advantages of insect meat compared with regular meat, and wondered why these advantages had remained misunderstood for so long. Indeed, we now know that producing insects is economically advantageous, leads to less environmental damage (in terms of greenhouse gas and ammonia emissions), holds fewer health risks for humans, and yields a product that is comparable in quality to chicken, pork, or beef. It therefore seems logical to eat insects, but our impression of insects as dirty, creepy, and "to be avoided" has sustained a barrier to innovation for a long time. Therein lies our challenge for the coming years: to discover the many advantages of insects and to overcome prejudices. The best way to do that is to put the spotlight on the importance of insects for life on Earth, and to let people taste how delicious they are. During the Insect Experience Festival in Wageningen in 2011, we could barely keep up with the demand for pancakes and Popsicles containing visible mealworms. Visitors were eager to taste them, and combining the insects with well-known foods worked well.

In the past few years, people have been asking us regularly where you can buy insects, and how to cook them. Now that the first suppliers are established, we thought it was high time for an insect cookbook. With this book in your hands, you can experiment to your heart's content. Without a doubt, many more new, delicious recipes will follow.

4. Stefan Gates, "Smoked Jellyfish: The Roast of Christmas Future," *New Scientist*, December 24, 2010.

Information Resources

The Laboratory of Entomology at Wageningen University, the Netherlands, conducts research on the biology of insects and the ecological relationship of insects with other organisms. In addition to fundamental research to elucidate how insects function in their environment, applied research is directed toward using insects for societal applications and preventing problems caused by insects in agriculture and in human and animal health. For background information regarding insect consumption—including a list of all insects that are known to be eaten, articles, and links to other Web sites—see www.wageningenur.nl/ent and click on "Edible insects."

The Food and Agriculture Organization of the United Nations (FAO) publishes interesting information about insect consumption worldwide. For guides, publications, and links to other Web sites, see www.fao.org/forestry/edibleinsects/en/.

The International Insect Centre (www.insectccntre.com) regroups all actors in the value chain dealing with insects as food and feed.

Margje Calis carrying packaged freeze-dried insects. (Lotte Stekelenburg)

The book *Edible Insects: Future Prospects for Food and Feed Security* can be downloaded at www.fao.org/docrep/018/i3253e/i3253e00.htm.

For the *Food Insects Newsletter*, see http://www.foodinsectsnewsletter.org.

Venik (Dutch Insect Farmers Association [www.venik.nl]), the first organization of insect farms in the world, introduced insect farming to the Western world.

Suppliers of Insects for Human Consumption

Cricket bars are available in the United States from Chapul (www.chapul.com) and Chirp Farms (www.chirpfarms.com).

Insects farmed for human consumption in the Netherlands (Bugs Organic Food) can be obtained from Ruig and Sons at Bugs Originals (www.insects4food.org). In the United States, edible insects are available from World Entomophagy (www.worldentomophagy.com).

Weaver ant pupae at a market in Laos. (Arnold van Huis)

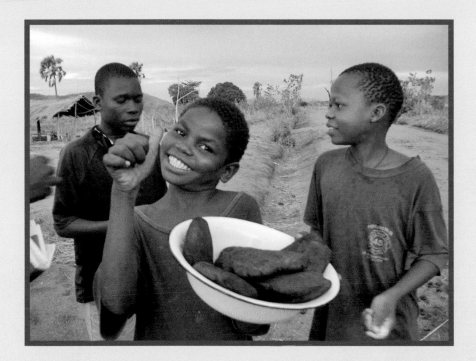

Boy carrying a bowl of *kungu* cakes in Malawi. (Midori Yajima)

Index

Numbers in italics refer to pages on which illustrations appear.

mealworms (*continued*)
122, 123; Chili con Carne, *96*, 97;
Chop Suey, *98*, 99; Flower Power
Salad, *70*, 71; fresh food needs of,
56; greenhouse gas production of,
169; Hakuna Matata, *94*, 95; in ice
cream, *17*; Insect Burgers, *102*, 103;
Jambalaya, *100*, 101; land required
for production of, *170*; Minestrone,
86, 87; Nutty Mealworms, 61, *61*;
in pesto, 39–40; Pizza, *126*, 127;
in *poffertjes*, 39; Pumpkin Soup,
76, 77; Quiche, *106*, 107; Quick
Meatballs, 62, *62*; Ravioli, 90–91, *91*;
Tagliatelle with Creamy Herb Sauce,
88, 89; Tarte Tatin, *148*, 149; yellow
mealworms (*Tenebrio molitor*), xv,
12–13, 55
meat alternatives, 168–70
meatballs, 60; Quick Meatballs, 62, *62*
Meat the Truth, 58
medication coating, 166
Meertens, Ruud, 55, 58, 59, 65, 82
Menzel, Peter, 111
Messengers of the Death installation
(Jan Fabre), *163*
methane, 168
Metius, Dirck, *167*
Mexican Chapulines, *42*, 43
Mexican grasshoppers *(chapulines)*,
34, *34*, 109
Mexico, 33; spirited caterpillars in, 35
mezcal drink, 35, *35*
migratory locusts (*Locusta migratoria*),
xv, 12, 55, 56, 67. *See also*
grasshoppers
milk production, 170, *170*
mineral content, 67, 68
Minestrone, *86*, 87
Mini Spring Rolls, *50*, 51
Mirror Room, Royal Palace of Belgium,
163, *164*
mopane bread, 167
mopane caterpillars: Mopane Caterpillar
Stew, 158; in southern Africa, 156–58,
157, *158*
moths: Bogong moth (*Agrotis infusa*),
143; in Italy and Australia, 142–43,
143; Savory Silkworm Cookies, 161;

silk moth larvae, 161; silk moth pupae
in China, 158–60, *159*, *160*; wax moth
larvae, 109–10, 112; *Zygaena* moths,
142, *143*
motivation, 114
mouthparts, 2
mushrooms: Wild Mushroom Risotto,
92, 93

natural insect dye, 116, 117
Naturalis Historia (Pliny the Elder), 159
Natural Red 4 food coloring, 116, *116*,
117
Natural Red E120 food coloring, *116*, 117
Netherlands, 33–34, 172; HAS University
of Applied Sciences, Den Bosch, 16;
pastry chef in, 113–15. *See also* Venik
Netherlands Food and Consumer
Product Safety Authority (NVWA), 55
"A New Episode in the History of Our
Civilization" (Wijffels), 171–74
New Scientist, 178
Niger, grasshoppers in, *37*
Nigeria, palm weevil larvae in, 118
nitrous oxide, 168
Noma (restaurant), 132, 137
Nordic Food Lab, 132
nutrition, 10, 17–18, 66–68
Nutty Mealworms, 61, *61*
NVWA. *See* Netherlands Food and
Consumer Product Safety Authority

Oatmeal Bars, 63, *63*
Oecophylla. See weaver ants
Opuntia cactus. *See* paddle cactus
overshoot, 171

packaging, 140
paddle cactus (*Opuntia* cactus), 116,
116
palm beetles, 67; larvae, 118, *118*, *119*
palm trees, 118
palm weevil larvae, 67, 118
pancakes, 39. *See also* Crêpes
Party for the Animals, 58
pasta: Ravioli, 90–91, *91*; Tagliatelle with
Creamy Herb Sauce, *88*, 89
perishability, 13
Peru, cochineal from, *116*, 116–17

WHO. *See* World Health Organization
Wijffels, Herman, 171–74
Wild Mushroom Risotto, *92*, 93
Willem van Loon and His Family as Israelites Gathering Manna (Metius), *167*
Wind, Pierre, 15–20
World Health Organization (WHO), 173

Xi Ling-Shi (Leizu), 158

Yang Yunan, 161
yellow jackets. *See* wasp larvae
yellow mealworms (*Tenebrio molitor*), xv, 12–13, 55. *See also* buffalo worms

zinc deficiency, 67
Zonocerus variegatus (stinking grasshopper), 64, *64*
Z@ppLive, 15, 17
Zygaena moths, 142, *143*

Arts and Traditions of the Table: Perspectives on Culinary History
Albert Sonnenfeld, Series Editor

Salt: Grain of Life, Pierre Laszlo, translated by Mary Beth Mader

Culture of the Fork, Giovanni Rebora, translated by Albert Sonnenfeld

French Gastronomy: The History and Geography of a Passion, Jean-Robert Pitte,
    translated by Jody Gladding

Pasta: The Story of a Universal Food, Silvano Serventi and Françoise Sabban,
    translated by Antony Shugar

Slow Food: The Case for Taste, Carlo Petrini, translated by William McCuaig

Italian Cuisine: A Cultural History, Alberto Capatti and Massimo Montanari,
    translated by Áine O'Healy

British Food: An Extraordinary Thousand Years of History, Colin Spencer

A Revolution in Eating: How the Quest for Food Shaped America, James E.
    McWilliams

Sacred Cow, Mad Cow: A History of Food Fears, Madeleine Ferrières,
    translated by Jody Gladding

Molecular Gastronomy: Exploring the Science of Flavor, Hervé This,
    translated by M. B. DeBevoise

Food Is Culture, Massimo Montanari, translated by Albert Sonnenfeld

Kitchen Mysteries: Revealing the Science of Cooking, Hervé This,
    translated by Jody Gladding

Hog and Hominy: Soul Food from Africa to America, Frederick Douglass Opie

Gastropolis: Food and New York City, edited by Annie Hauck-Lawson
    and Jonathan Deutsch

Building a Meal: From Molecular Gastronomy to Culinary Constructivism,
    Hervé This, translated by M. B. DeBevoise

Eating History: Thirty Turning Points in the Making of American Cuisine,
    Andrew F. Smith

The Science of the Oven, Hervé This, translated by Jody Gladding

Pomodoro! A History of the Tomato in Italy, David Gentilcore

Cheese, Pears, and History in a Proverb, Massimo Montanari,
    translated by Beth Archer Brombert

Food and Faith in Christian Culture, edited by Ken Albala and Trudy Eden

The Kitchen as Laboratory: Reflections on the Science of Food and Cooking,
    edited by César Vega, Job Ubbink, and Erik van der Linden

Creamy and Crunchy: An Informal History of Peanut Butter, the All-American Food,
    Jon Krampner

Let the Meatballs Rest: And Other Stories About Food and Culture,
    Massimo Montanari, translated by Beth Archer Brombert

The Secret Financial Life of Food: From Commodities Markets to Supermarkets,
    Kara Newman

Drinking History: Fifteen Turning Points in the Making of American Beverages,
    Andrew Smith

Italian Identity in the Kitchen, or Food and the Nation, Massimo Montanari,
    translated by Beth Archer Brombert

Fashioning Appetite: Restaurants and the Making of Modern Identity,
    Joanne Finkelstein

The Land of the Five Flavors: A Cultural History of Chinese Cuisine,
    Thomas O. Höllmann, translated by Karen Margolis